50 Years
in the Semiconductor
Underground

The author (left), when he was much younger, prepares to join a colleague in the underground salt mines near Salzburg.

50 Years
in the Semiconductor
Underground

David K. Ferry

PAN STANFORD PUBLISHING

Published by

Pan Stanford Publishing Pte. Ltd.
Penthouse Level, Suntec Tower 3
8 Temasek Boulevard
Singapore 038988

Email: editorial@panstanford.com
Web: www.panstanford.com

British Library Cataloguing-in-Publication Data
A catalogue record for this book is available from the British Library.

50 Years in the Semiconductor Underground

ISBN 978-981-4613-34-7 (Hardcover)
ISBN 978-981-4613-35-4 (eBook)

Printed in the USA

Contents

Preface

This book has grown out of a talk I gave at the International Conference on Advanced Nanodevices and Nanotechnology in Hawaii a few years ago. For some reason, the conference was dedicated to me, and so I had to give the introductory lecture, which more or less covered a few of my contributions to the field. The talk was titled the same as this book, and I tried to have a little fun with it. Now, there are two important points about this talk and this book. First, one normally doesn't dedicate a conference to a living person, but this particular conference has made a habit of doing so. But, in the end, the person did cease to exist within a few months to a few years after the conference, and so never made it back to the conference. The first person honored was Rolf Landauer from IBM in 1998, and he died just a few months later. Rolf was a pioneer in statistical physics and the world we now call mesoscopic or nanoelectronics. Then, they honored Prof. Susumu Namba from Osaka, and he died a couple of years later. Susumu was very important in the development of ion beam processing and again in mesoscopic nanoelectronics. Following this, the conference honored Prof. Gottfried Landwehr from Würzburg (Germany), and he never made it back to Hawaii, finally dying in 2013. Gottfried was quite important in bringing together the world of semiconductor physics in high magnetic fields, and founded the conferences on this topic. Finally, they honored me, I suppose for my longevity in the field. Considering the history of previous honorees, I made a point of saying that it seemed in reality that they were just trying to get rid of me.

But, if you are reading this, it is likely that they have failed and I have returned once again to criticize some of the goings-on of the conference.

The second point that is important is that one cannot talk for less than half an hour and cover the material that makes it into a book like this. Indeed, even this book cannot contain everything, and would likely be boring if it did. The intervening time from the conference to beginning this book was filled at first with doubt about such a book. But then I started thinking about the various confusions and other such things that have occurred in science, as well as my life, in the approximately 50 years that I have actively pursued the issue. It was clear that there were enough interesting events that conceivably could make such a book possible. And then there were the urgings from the publisher, who seemed to think that it would be worthwhile to do the book. So, that is the task I undertook, the results of which fill the following pages. It should be remarked, however, that I have been known to go off on long rants, which if written down and put together clearly would also fill a book. Certainly, some of these appear within the topics that are discussed, and I leave it to the reader to decide which is which.

In some sense, as I write the book, I feel not unlike the aged samurai in *A Book of Five Rings* (Go Rin No Sho) as he sits in his cave writing his memoirs. But his book was rules and advice to be followed by a young samurai. This one is anything but. Instead, this is a book about random encounters and my views on a number of topics. This book is *about* science and engineering, but is not *on* science and engineering. It is not a textbook which develops the understanding of a small part of the field. It is really a book about the strengths and the foibles of living as a physicist/engineer for half a century. I should point out that in no way is this book anything close to an autobiography. Such a thing would quite likely be very boring. Rather, the book presents my personal views about science, engineering, and life, illustrated by a number of stories about various events, some of which have shaped my life. In the telling of these to my colleagues, some have been humorous, so perhaps that will come through the words that make this book. And sometimes, the claims of others, which have stretched the meaning of reality, have seemed so ridiculous as to encourage a bit of laughter. Some of these incidents will also appear, but I will try not to name names where that might demean the participants.

Many friends and colleagues have asked why haven't I retired and enjoyed life, since I am still very active and un-retired. There are a couple of reasons that I remain so. One is that my wife has refused to let me retire until I give her a "plan." This plan should give details of how I intend to spend my days after retirement. Since, I don't have such a plan, I continue to carry on. Secondly, of course, I still enjoy both research and warping the minds of young people (this is called teaching). In a large sense, I am still enjoying myself too much to consider giving it up. I suppose I could retire and play golf, but I gave up golf many years ago when I ran out of clubs. Such a game, which generates so much frustration, holds no great allure to me now. So, perhaps more events will occur in the future and add to my trove of interesting (at least to me) stories.

Thanks go to a great many people who have influenced me through the years, some of whom continue to do so. Not the least is my wife (of more than 50 years) and my family, who have put up with me since the beginning. Then, there are my students, postdocs, and colleagues who have survived my rants and tirades and still have managed to become very competent individuals. Some of them are mentioned, but none of the book would have been possible without them. They should all know that I was thinking of them as I wrote the words, because they made it possible. There have been a few individuals who have made significant contributions to what has become my career that I do need to thank. These include Jon Bird, Steve Goodnick, and Richard Akis in the US; John Barker, Carlo Jacoboni, and Antti Jauho in Europe; and Chihiro Hamaguchi, Koji Ishibashi, and Yuichi Ochiai in Japan. Very special thanks must also go to Larry Cooper, who was a colleague at the Office of Naval Research, and whose support over the years has made much of my work possible. Finally, I would also like to thank Alex Kirk for reading the manuscript and pointing out my flaws. Finally, thanks to Stanford Chong, who encouraged me to consider doing this project.

Dave Ferry

Chapter 1

In the Beginning

Growing up in south-central Texas may have been as good a beginning as any other. Texas is an interesting place, and life there is influenced by a great many trends from the past. A key facet of this is that Texans remain proud of the fact that the place was an independent country that chose to join the United States. That independence of spirit continues to carry forward within the citizens. More importantly, Texas is a central place where the old south mixes with the wild west to produce a somewhat unique blend (or perhaps its paranoia) within a person. The state is divided by a line which roughly runs along the Balcones Fault stretching from a point midway between Dallas and Fort Worth, south through Austin and San Antonio. To the east lies what is called "black dirt" country, where there are forests and heavy rainfall. This area is an extension of the old south. However, to the west of this line lies "red dirt" country and is arid grassland. This latter is West Texas, which is very different from East Texas. West Texas is the land of the plains Indians. Growing up in San Antonio meant being influenced by both sides of this great divide, or perhaps it explains the general confusion

50 Years in the Semiconductor Underground
David K. Ferry
Copyright © 2015 Pan Stanford Publishing Pte. Ltd.
ISBN 978-981-4613-34-7 (Hardcover), 978-981-4613-35-4 (eBook)
www.panstanford.com

with often permeates my thinking. Nevertheless, it provided an environment in which my views of life were formulated and it shaped the directions I would take.

It appears that a series of events, both large and small, lead a person to their adopted vocation. The effect of each of these events differs depending upon the environment in which they occur. To one person they may be life changing, while to someone else it may be just another stop at the fast food joint. Thus, the creative events that lead one down a particular garden path are unique in the way they affect the particular individual within his or her environment. For me, this sets up the question "why semiconductors?"

Why Semiconductors

The transistor was invented at Bell Telephone Laboratories in 1947. But, like the rest of the world, it held no meaning for to me until it impacted my life. In the mid-1950s, the CK722 was introduced as the first low-cost transistor available to the general public (ca. 1954). I was involved with an amateur radio club in high school, and this became an exciting new event. One could actually buy a kit to produce a small personal radio. It had a socket in which you could plug in a set of headphones, but it also used a long wire as an antenna. With one hand you swung the antenna over your head and with the other you tried to tune the radio. This actually worked on occasion. Progress was swift. Within a year, a commercial transistor radio, with built-in antenna and speaker, was on the market (there is some debate about who achieved this first, as several appeared at about the same time). It was absolutely clear that a major change in the electronics world was under way.

These new gadgets, the transistors, were made of semiconductors, of which we knew little. As a material, they had been around for more than a century, and a historical search showed that they had been used in microwave detectors for almost as long. But, in our limited world of school electronics, there was no evidence for their existence. Still, these miniature radios were amazing, and one could believe that we would have Dick Tracy's wrist radio within a few years. These were the precursors to the beginning of the information age that followed from the appearance of the integrated circuit at the end of the decade, an event we return to in a later chapter. This decisive trigger enhanced the thrill and interest in electronics and set me on my future in electrical engineering.

As a field, electrical engineering is very broad, basically addressing almost every aspect of all technology connected in any way to electricity. But semiconductors were not a big part of the curriculum to which I was subjected. Other factors had to appear to keep me dedicated to pursuing my desired topic. Many of these were negative factors, not the least of which was my introduction to rotating machines. As part of the curriculum, we were subjected to laboratory courses in both d.c. and a.c. motors. But, these were not the wimpy little desktop machines that became popular in schools many years later. These were real-world, massive machines. The d.c. motor that provided a memorable experience was about a foot and a half in diameter and somewhat larger in length. We were supposed to measure the speed of the motor as a function of the current sent into the "field" coils that produced the magnetic field that was necessary for the motor to operate. To get the required d.c. power, we had to use a large a.c. motor/generator pair. Wiring was done with massive wires that seemed to be larger than my fingers. These were equipped with humongous plugs, which were inserted into a giant power panel. The entire system looked like something out of a horror movie (and it certainly seemed that way to me at the time). Now, this particular d.c. motor, which we were allegedly characterizing, was a type in which the speed was inversely proportional to the current in the field coils, which itself was separate from the excitation current to the rotating part of the motor. One of the important laws of engineering is Murphy's law, which has a great many descriptions. Usually, it means that disaster is just about to happen. In the present context, Murphy's law meant that if anything could go wrong, it would. Sure enough, my lab partner tripped over the wires, which resulted in disconnecting the drive to the field coils. This sent the motor speed skyward, resulting in a great vibration which rattled the entire building. I managed to disconnect the main power before disaster arrived. Nevertheless, the professor came running into the lab, shouting, "What are you doing?"

"The lab, sir," I responded. Obviously, this was a correct response to his ill-conditioned question, but it was not at all clear that he approved of this answer. We will have some more to say later about ill-conditioned questions, but at the moment that was not a point upon which I was spending much thought process. It was not at all clear to me that I still had a future in engineering, thanks to my hapless lab partner. But most sins are readily forgiven, and that seemed to be the case here, at least for the moment.

With a new semester came a new power laboratory, this time with the a.c. motor. This new machine was considerably larger than the d.c. equivalent, and was a three-phase machine—a real beast. Our project of the day was to measure the torque produced as a function of the excitation currents. This involved bolting a massive wooden beam to the shaft of the motor. My recollection is that this beam was a 2×8, or something of that ilk. The idea is that one end of the beam was bolted to the shaft and the other end was restrained by a really large spring attached to a scale. This would restrain the motor, and as it tried to rotate it would pull against the spring and scale. Measuring the force on the scale and knowing the length of the beam, you could determine the torque. Now, three-phase electricity is kind of interesting, which is a polite way to express one's view of it. If you label the three phases as A, B, and C (rather unique it seems), then the sinusoidal voltage of phase A would peak at one point in time. Phase B would then peak 1/3 of a period later, and phase C still another 1/3 of a period later. So, if the motor was excited with phase sequence ABCABC . . ., it would rotate one direction. But if you screwed up and connected it as ACBACB . . ., it would rotate the opposite direction. Sure enough, when we energized the motor, the beam started in the wrong direction, ripping itself free from the spring and rotating rapidly over the top of the motor toward where we were standing. While we managed to avoid the beam, it crashed into the floor with a significant boom. Once more, the professor raced into the room, shouting, "What are you doing?"

"The lab, sir," I responded. As before, he seemed incredulous at our behavior. Now, I was sure my future was doomed. But, in the end, all was forgiven and we could move on.

However, seeing a massive wooden beam coming at you from above is a marvelous method of focusing one's thought processes. It was absolutely clear at that moment that my future was not supposed to be in the electric power field. There was no way that I wanted to mess with such big machines ever again. Instead, I remembered those tiny transistors and these events refocused me into semiconductors as a road to the future.

Many years later, these events came back to me in a positive way. At the time, I was the director of our Center for Solid State Electronics Research, and we were still shaking down a new cleanroom. We were meeting with university staff and contractors to make final changes, and I asked them to please check the phase sequencing of the three-phase motors on the exhaust fans for our chemical hoods. These fans were located on the roof, where we really didn't have access to them. The question was asked, because we didn't seem to have a very powerful exhaust in our chemical hoods. Sure enough, they were phased wrong, and the fans were blowing down the stack instead of pulling air up it. The only reason we had any net exhaust was the high overpressure that was maintained in the cleanroom. The group thought I was a genius, but I had merely remembered my own unhappy experience with three-phase motors.

So, my life seemed headed toward semiconductors. When I started graduate school, the curriculum had finally added a course on transistors. Of course, there was no one on the faculty who could teach this course, which meant that some non-expert had to fill in. This, in turn, meant that it was unlikely that a rational answer to any deep question would be forthcoming. As a result, understanding came relatively slowly. But it seemed that there was an opening here, as it was probably true that there weren't many experts anywhere in the universities. So, I persevered as much from interest as from the realization that they probably wouldn't bash me on the top of the head. Fortunately, there was plenty of information available (even without the Internet at that time), and a voracious reading program came to solve most ills.

The Underground

One might wonder about just what is meant by the word underground in the title. In the ancient world (anything more than half a century old for the present purpose), society was run by the ruling elite. The rest of the world was faced with a somewhat less enjoyable existence. There were levels, or classes, of subservience, and a large fraction spent their time toiling in the mines, or the underground. It didn't really matter whether it was a real mine or a spiritual or perceptional mine. The workers had the feeling that they were grinding away in the underground of society. I did my doctoral work at the University of Texas (in Austin, as that was basically the only campus at that time). In later years, my professor at Texas would visit my former school and talk about how (financially) poor Texas was. My former professors would challenge how someone from the University of Texas could consider themselves poor. He would then explain: "Yes, the University of Texas is poor, but you guys are poverty stricken." Fortuitously for me, the University of Texas recently had invested heavily in upgrading their graduate research programs, bringing in many new faculty, and my professor had moved there from the University of Illinois. I would say that this was probably the first round in which schools not in the elite upper class realized the need to upgrade their quality. Even then, it was recognized that this could be an economic driving force for the state. Nevertheless, no one would be mentioning Texas in the same breath as Harvard or Stanford at that time.

In science and engineering today, that class distinction remains. There are the elite institutions, a few up-and-comers like Texas, and the rest of us, who may be described as wannabes. Everyone proclaims schools such as Harvard, Yale, Princeton, Berkeley, Caltech, Stanford, and MIT as the so-called elite. If one doubts this, just ask the residents of these institutions. People who do surveys, such as *U.S. News and World Reports*, always tell us that these are the elite. Up until a couple of decades ago, you could add corporate research labs, such as Bell Telephone Labs, to this list, but their fortunes have greatly declined since the breakup of the company. So, in the society of scientists and engineers, these groups are the royalty, those from the up-and-comers are the knights and generals, and the rest of us, the wannabes and lesser schools, are doomed to wander through life as a soldier or serf.

Now, one wonders how much of this is a self-serving image that is propagated from above. For sure, the number of such surveys and evaluations has exponentially increased in recent years, just as have the announcements that these surveys really matter. Could it be that these surveys are driven by alumni from these same institutions as a means to reinforce their satisfaction at having spent so much money for an education from one of these same institutions?

Certainly, there exists a bias in many circles against the non-elite institutions. And this makes it more difficult to sustain any attempts at upward mobility among the lower-level institutions. I remember a discussion in the '70s with a program manager at the National Science Foundation who stated that peer review meant that your proposal would be evaluated from alleged peers at equivalent universities. Thus, if you were proposing research at the same level as say MIT, but were from a midwestern school, it is quite possible that your proposal would be reviewed not by an expert in the particular scientific area but by someone at another mid-something or other university who really knew little about the field. This would certainly produce a review that would not be to your best advantage. This persecution continued until Congress mandated that the agency spend its money more uniformly across the country. This does not seem to have hurt the science output of the country. On the contrary, it seems to have helped produce many new leaders from places other than the elite.

One might be driven to ask a question as to whether the elite really deserve that status. But it is clear that the elite make a great effort to attract the most outstanding scientists to their faculties. Of course, other schools try to do the same thing, but as you go down the hierarchy, the results become more and more patchy. One might ask whether this performance can be measured. But this brings us back to the surveys. As discussed, these are often self-serving, so we should seek another measure.

If we look at the most famous scientists, then the answer is not so helpful. Certainly, Einstein didn't attend one of these schools, nor even an elite German school. Rather, he went off to Switzerland to attend university, and here his record was not all that good. Upon graduating, the best job opportunity was sorting patents. History holds him in a somewhat higher status level, where he is regarded as perhaps the most brilliant scientist of the twentieth century. The only person who has won two Nobel prizes in physics is John Bardeen, who did his undergraduate degree in electrical engineering at the University of Wisconsin, and then did his graduate work at the University of Minnesota. While he won his first prize for work at Bell Labs, he had already left for the University of Illinois, where his work led to the second prize. So far, no elite schools here.

What about the formulative change that has led the last 50 years to be called the information age? That is, what about the world of computers, microchips, and the Internet? Well, one hears about computers arising on the east coast, but that is the publicity of the thing. The idea of a mechanical computing machine dates to Charles Babbage in the mid-nineteenth century, but it took more than a century before one could actually make a complete working version of his concepts. Certainly, World War II brought the digital age to fruition with electronic machines developed in or at the end of the war in the US, the UK, and Germany, many under government support. But, in 1972, the courts ruled that the first electronic computer was the Atanasoff-Berry Computer, developed at Iowa State University and first operated around 1942.* Atanasoff received his electrical engineering degree from Florida, and followed with graduate degrees from Iowa State and Wisconsin. Berry was an electrical engineering student who finished at Iowa State.

The integrated circuit came to us from Jack Kilby at Texas Instruments and Robert Noyce at Fairchild Semiconductor. Their inventions were independent, but they started the growth of computing power that we now enjoy. Kilby received his electrical engineering degree from the University of Illinois, and a master's degree from the University of Wisconsin—Milwaukee. Noyce did his undergraduate at small Grinnell College, but did go on to MIT for his doctorate. Finally, a degree from one of the elite. We defer the full story of Kilby to a later chapter, but he ultimately won the Nobel Prize for this work (Noyce had died before the prize was awarded; Jack accepted in his name as well).

The Internet arose from government supported research in the mid-'60s. The generally acknowledged start was the sending of a message from UCLA to Stanford Research Institute in Palo Alto, involving Leonard Kleinrock and Doug Engelbart (of UCLA and SRI, respectively). Kleinrock did his undergraduate work at City College of New York and graduate work at MIT,

---

*http://en.wikipedia.org/wiki/Atanasoff–Berry_Computer

while Engelbart did his undergraduate at Oregon State and graduate degree at Berkeley. Now, here are two more graduate degrees from the elite.

So, if we review these significant events from science and engineering, we find that there is a presence from the elite, but they do not dominate the history to the extent that many would have us believe. In fact, we can go a little further. A couple of years ago, the *Wall Street Journal* asked the major manufacturing corporations from what schools they preferred to hire their engineers. The results were quite amazing, in that no elite schools made the top parts of the list. Instead, the top school was Penn State, and Arizona State placed fifth. Indeed the top half of the list was large state universities. But this result might just be due to the fact that large state universities have significantly more graduates than the elite schools.

The point of this discussion was not to demean the elite, but to illustrate the fact that there is a very small difference between the elite schools and a large number of American universities across this country. However, one can certainly accuse me of the same bias as I mentioned above and perhaps my selection of events is not representative. OK, this is probably true. In fact, to counter such a bias, one can consider the Nobel prizes in physics as some kind of guideline (certainly as useful an evaluation as that of *U.S. News and World Report*). By my count, in the 50 years from 1963 to 2012, 115 different individuals won or shared the Nobel Prize in physics. Of these, some 58 were entirely educated outside the US. Of the remainder, who received some education in the US, 18 received a bachelor's degree and 32 received a doctorate from the elite institutions listed earlier. That is, more than half received their doctorate from one of these seven institutions. On the other hand, it also means that almost half received their doctorate from another school, and that these other schools may well be just about as good. From the stories above, it is clear that the elite schools have produced greatness, but a great many scientific and engineering achievements have arisen from these other schools as well.

While it may be argued that artificial creation of various levels of status focus most likely upon the wrong measures, awards for the best and brightest suggest a realistic elite education may exist. Are the elite really better? Who knows, and how the question is asked may be more important (a point we return to later). In fact, a recent evaluation by a Chinese university has put Harvard, Berkeley, Stanford, MIT, and Caltech as five of the top six universities in the world (only Cambridge University, at number 5, broke the US grip).* The elite do a good job. Nevertheless, they leave many of us with the feeling of working in the underground!

The Joy of Semiconductors

Semiconductors are a ubiquitous material in our everyday life. Nearly all semiconductors have a crystal structure very similar to that of diamond, but differ slightly in the spacing of the atoms and in the details of the atomic potentials. These subtle differences give rise to a great variety of properties. But, in nearly all cases, the atoms have (on average) four outer-shell electrons. In chemistry, these outer-shell electrons are often referred to as the valence electrons. In our semiconductors, these outer-shell electrons form the chemical bonds which hold the crystal together. And, in semiconductors, there are just enough outer-shell electrons to fill all possible states in the valence band formed by the bonding process. The next available energy levels lie in a higher band, which is referred to as the conduction band. So, in normal circumstances, the semiconductor would be a weak insulator in that high temperatures can cause an electron to jump over the gap between the valence band and the conduction band. But, in the semiconductor technology, this weak insulator behavior gives us a blank canvas upon which to create electron devices either singly or in the billions for a microchip such as a microprocessor. Each electron device, or transistor, is formed by

---

*http://www.shanghairanking.com/ARWU2013.html

the controlled introduction of impurities and by the formation of small structures via multiple layers of growth, deposition, etching, and photolithography. The finished microprocessor is a manufacturing marvel.

To understand the semiconductor and the electron device requires more than a simple understanding of classical physics. The electrons move in the crystal by quantum mechanical processes, and it is these that lead to the chemical bonding, the nature of the valence and conduction bands, and the manner in which the electrons respond to the external voltages and fields which are applied. Hence, each electron device is a quantum object in its own right and its operation is defined by a great many aspects of modern physics.

But semiconductors are more than just microprocessors and individual transistors. Semiconductor light sensors have become even more ubiquitous as they have replaced other methods of light sensing in detectors and solar cells. The reverse process has led to light-emitting diodes (LEDs) and semiconductor lasers. And arrays of light sensors, beginning with charge coupled devices several decades ago, have eliminated film from the world of cameras, with digital imaging (and resulting processing to improve the image) becoming the norm. LEDs are revolutionizing the lighting industry, as well as the standard model of a high-definition television. The microprocessor, which has revolutionized computing, has also taken standard old devices and modernized them and made them smart, from the telephone to the automobile.

Most semiconductors are weakly piezoelectric, in that stress can lead to induced electric fields, and vice versa. Hence, semiconductor cantilevers can be used to sense motion via the acceleration-induced bending of the cantilevers. This has led to new application fields in microelectromechanical (or MEMS) systems. Among the first applications in the real world were as sensors for the deployment of auto airbags, because such semiconductor sensors are easily integrated with the other microelectronics that generate the control and activation signals. But one encounters these every time one rotates or turns one's tablet or smart phone, as the rotating motion must be sensed to tell the device to rotate the display for you.

Indeed, it is the growth of the integrated chip over the past 50+ years that has enabled the information age to develop, and it is this growth that has revolutionized the way we interface with life today. Fifty years ago, we were excited by the introduction of push-button phones doing away with the old rotary dials. Today, the smart phone can be used to control or affect almost every aspect of your daily life. One can say that, from the single transistor and rough start of my youth, semiconductors have become the magic material upon which much of our life today depends.

My Life and Times

As I pointed out in the Preface, this book is not an autobiography. Nor is it a textbook on semiconductors or integrated circuits. It is a description of a number of events and/or experiences which have affected my life beginning with those mentioned above which led me to the field of semiconductors. I will try to join similar concepts (and adventures) into a single chapter, but this is not always possible or even desirable. And the topics will stray from simply semiconductors because the study and understanding of semiconductors depends upon many fields of science, as mentioned above. For example, quantum mechanics has a big impact upon semiconductors, but the rise of the integrated circuits surely has an impact upon the interpretation of quantum mechanics. As there are several possible (and widely discussed) interpretations, we will have to have some discussion of what quantum mechanics means in this field of interest.

Importantly, this is not a book on the history of semiconductors, so I do not feel compelled to provide an exhaustive bibliography on the topics which will be discussed.

In some cases, I may give a citation to some key papers, and, of course, I will provide the source when I use direct quotes or ideas from other people. Most importantly, the discussion will focus on my views of a topic and my impression of the events under discussion. If one wants to delve deeper into a particular topic, more information can always be found on the Net, which itself is another manifestation of the way in which the information age has changed our life. I continue to enjoy the life I lead, and I hope that you will find the remaining material of interest, and perhaps even controversial at times.

Chapter 2

Threads of Science

In 1978, James Burke published a book called *Connections* (it has since been reissued), and he also did a series on PBS with the same title. The purpose was to identify a thread that connected a number of discoveries and advances in science and technology over a long period of time. In discussing a particular thread as it wove through time, he followed what seemed to him to be a path in the development of man. Each path was focused on a single theme, such as the importance of metallurgy. While most historians of science would probably have different views on such developments than Burke, his work does illustrate the fact that independent events in science sometimes are coupled over the years by such developments. It is certainly clear that Max Planck's work on blackbody radiation at the end of the nineteenth century led him to postulate that light was actually composed of small particle-like entities. That is, light was quantized into units of energy. This idea then led Einstein, in 1905, to develop an explanation of photoemission solely in terms of particle properties and the classical ideas of conservation of energy and momentum. The conundrum of the early twentieth

50 Years in the Semiconductor Underground
David K. Ferry
Copyright © 2015 Pan Stanford Publishing Pte. Ltd.
ISBN 978-981-4613-34-7 (Hardcover), 978-981-4613-35-4 (eBook)
www.panstanford.com

century also fed other events. Niels Bohr developed his model of the atom based upon each electron having a well-established orbit around the nucleus with the proviso that such orbits would not exhibit the normally expected decay of amplitude. The rest of the theory was basically classical physics. The Bohr theory led Louis de Broglie to suggest that the stability of such orbits lay in the fact that the electrons actually were waves. Hence, each orbit existed with a size demand that allowed only a fixed number of wavelengths in the circumference. This interplay between waves and particles ultimately led to the development of modern quantum mechanics by Heisenberg and Schrödinger, to name just the two most popular. There were certainly many others who had a hand in the development of the field.

But there are other threads that wind their way through life. Who among us today ever considers that their quarter acre of land was a concept that has its basis with the Romans? Or who would think that the obscure unit from horse racing—the furlong—is involved in this history? When I was young, I developed a very good understanding of just how large a block was. As a normal child, I often got on my parents' nerves, at which point my father would tell me to "go ride your tricycle around the block." Of course, after doing so, I would then ask, "Now what?" As I grew, I learned that eight blocks made a mile, although my understanding of what a mile really meant was rather obscure. Now, you may ask, "Why eight blocks to a mile?" And that is where we connect with the historical mile. It is fair to ask just why we should worry about this in the discussion of a life in semiconductors. It turns out, however, that the results of history impact us today, and the end result is very useful in describing a number of topics I want to discuss. That is, the layout of a modern city is seen time and again in science, and particularly in semiconductors and microprocessors.

History tells us that our word *mile* comes from the Latin *mille*, which actually means 1000. The Romans would mark distance in terms of 1000 paces, each of which was two steps—one left,

A Roman *mille* marker (on the right) in Mauterndorf, Austria. This small town is located on the Roman route over the Alps via the Tauern pass.

one right—as marched off by their armies. Often, a marker was placed at each mile point, and some of these have survived to the present. So, wherever the Romans marched, locals developed their own interpretation of this distance. As one may expect, this led to confusion between different interpretations of the distance. But, under Elizabeth I, when the English were planning their conquests, parliament provided us with the *statute mile*, 5280 feet or eight furlongs long. (Each furlong was 40 *poles*, each of which was 16.5 feet.) Now you know: we have eight blocks, because each block is a furlong in length. While the official international mile is different by a miniscule amount, the statute mile is used often in land surveys. In some sense, this is an advantage when it comes to dimensions smaller than the mile. Almost anyone can divide a length into two nearly equal parts. If I repeat this twice, I have the eighths needed for the block. Hence, each square mile is easily divided into 64 units of 10 acres each. See how easy land division becomes. Try this with metric units where you need to divide into 10 equal units. First, you have to divide into five nearly equal parts, which is very difficult to do without a complicated geometrical arrangement. The English way is better for this purpose. As a consequence, most Western cities, surveyed by the US geological survey, are laid out in square mile blocks, and the streets dividing these blocks are the major thoroughfares. In Phoenix, for example, we have 7th Street (OK, we messed up on that one), 16th Street, 24th Street, and so on, for the major north-south roads east of central avenue. To the west, we have 7th Avenue, 16th Avenue, 24th Avenue, and so on. Why the subtle street versus avenue difference? I suppose that is bureaucracy at work. If you think that it can't get any worse, visit Salt Lake City.

Now, we use exactly this rectilinear behavior in laying out the devices and circuits on an integrated circuit chip. The smallest unit is the gate, usually made up of a pair of transistors. For a variety of manufacturing reasons, the gates are laid out like our blocks in the city. Collections of gates then become larger functional units such as registers, adders, and so on. Wiring interconnects, as well as gate edges, run only along the rectilinear directions, which we may as well call east-west and north-south. So, our arrangement of city blocks has an exact analogue with the arrangement of transistors in an integrated circuit. The interconnects are the streets and highways. We will encounter other places where this analogy will prove useful.

While I would like to say that this chapter will focus on such useful threads as they wend their way through the history of science, this would not be a true summary. This is the case for a number of reasons. For example, a thread may only exist for a few years and then die when it is finally realized that it just wasn't a good idea. Or the thread might not be a single idea, but a philosophy of engineering, like good design (or bad design). This latter often has an important corollary—the law of unintended consequences. History is full of various designs that overlooked one seemingly trivial point, with the resulting disastrous results. Nevertheless, I have been involved with a number of threads, some long and some short. So, here I discuss a few of these threads in this spirit.

HILLS AND VALLEYS

Just above, we discussed the idea of square mile divisions in the layout of streets in many American cities. Traffic can move quite rapidly on these major mile streets. But within the area enclosed by these streets, traffic moves much slower. We can think about these enclosed areas as being hills surrounded by the mile streets. On these hills, traffic is slowed immensely. A similar effect is found in the band structure of semiconductors. The main compound materials have a central valley, which is like the mile streets, and upper valleys, which are like the hills that are enclosed. Under proper conditions, such as created by a high electric field (high voltage), the electrons can be excited into the upper valleys, where they go much slower. It is this large slowdown of the movement which is crucial to the Gunn effect.

Another way of thinking about this is to consider the landscape of Switzerland. There are valleys through which the major highways pass, and where cars can go at a relatively high rate of speed. But there are high mountain valleys, where the livestock are taken for the summer, and where the traffic moves very slowly. This provides the exact same behavior as the mile streets and hills above.

My Thread

When I was pursuing my doctoral research, I began to work on indium antimonide (InSb), a narrow-bandgap material of interest to a variety of applications. When the material was cooled to liquid nitrogen temperature (77 K, −321 F), and a magnetic field and voltage were applied, the material emitted a broad range of microwave noise. I spent a great deal of time characterizing the noise as well as trying to determine just what caused the signal to exist within the physics of InSb transport. It would take many years to finally nail down the exact cause. A year earlier, John Gunn at IBM had discovered coherent microwave oscillations in gallium arsenide (GaAs). At the time, it was thought that the emission from InSb was different since this emission was not coherent, and the frequency was not related to a time determined by the transit of electrons at their saturated drift velocity.

The Gunn effect, as the coherent oscillations in GaAs came to be called, was later determined to be due to the transfer of electrons from the central valley of the conduction band to the satellite upper valleys of the conduction band (see the aside, if you are not familiar with the band structure of semiconductors). In the central valley, electrons can travel at very high speeds, perhaps a factor of 2–4 times larger than in the upper valleys. Hence, the mobility—the response of the electrons to the electric field—is much smaller in the upper valleys, and this leads to a decrease in the current going through the semiconductor. Thus, as the electric field is increased, the current increases as well up to a point where the transfer to the satellite valleys begins. Then, the current decreases for further increases in the electric field. This leads to a situation where the electric field becomes inhomogeneous. Where the electrons go slow, the field must be much higher in order to make the current through the device have a constant value throughout the entire device. Since the high electric field region moves with the moving electrons, they will reach the drain end where the electrons exit the device. Then the

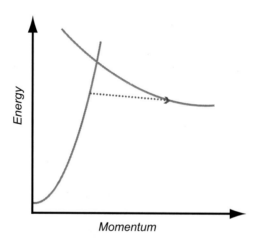

Conduction band of GaAs or InSb. Electrons normally reside in the lower band, but can scatter to the upper valley if they gain sufficient energy from the applied electric field. Such a scattering is shown by the dotted arrow.

process begins again with the high field region at the entrance of the device. This leads to an oscillation in which the transit time is given by the length of the device L and the upper valley velocity v as $T = L/v$. The reciprocal of this gives the frequency of the oscillation. Hence, the frequency is set by the length of the device. The velocity of importance is that of the slow electrons, as this is the region in which the high electric field exists.

Now, as I mentioned, it was generally believed that this effect could not be the cause of the wide-band noise generated in InSb. First, the frequency was not a single coherent value that could be related to the size of the device. Second, the bandgap in InSb was only about 0.2 eV, while it was thought that the satellite valleys lay some 0.4 eV above the minimum of the conduction band. Hence, it was thought that breakdown of the material, where electron–hole pairs are produced in massive numbers would occur at much lower electric fields than would be required for transfer to the upper valleys. In fact, measurements of the pair generation that led to breakdown clearly showed that this process would start at about 200 V/cm (at 77 K). So, it appeared that there was no way that the Gunn effect could be occurring in InSb.

However, working with colleagues from Vienna, and using very fast pulse techniques, we found that the process leading to pair formation was a very slow process, while the process that led to intervalley transfer was a fast process. Thus, if we looked on the nanosecond time scale, we could push electrons to electric fields above 600 V/cm before they could begin to produce the electron–hole pairs. Indeed, intervalley transfer could be seen to occur just above 600 V/cm. Hence, on the short time scale, or in very small devices, the Gunn effect could occur after all. But why then was the microwave emission so broadband, and not the single frequency that typified the Gunn effect?

It has long been known that one can create an electronic oscillator with transistors. But, in the presence of a load resistor, the oscillator does not perform as well as when it is left on its own. In fact, if the load gets too large, then the oscillations will be too heavily damped, and they break up into uncorrelated noise. Scientists at MIT's Lincoln Labs showed that they could use a resistor to load a Gunn oscillator to the point where the coherent oscillations also broke up into noise, and this suggested the mechanism which was occurring in InSb. In order to confirm this, we needed to probe deeply within the semiconductor. Following a similar idea from Gunn himself, we made a capacitive probe which could move along the edge of the semiconductor sample along its length. Sure enough, a high electric field region existed near the source end of the device, where the electrons entered. But the field dropped rapidly to a very low level within the rest of the device. So, the intervalley transfer only occurred very close to the source. The rest of the device acted like a huge load resistor and damped the oscillations. But we could detect the motion of the domains, and they moved with the upper valley velocity, just as in the Gunn effect. However, these oscillations were not coherent, and the large resistance of the remainder of the device damped them so that only wide-band noise resulted. It was the Gunn effect in the worst possible scenario. The role of the magnetic field was merely to lower the voltage required to start the process. This occurred through a geometrical effect and not from the physics.

Many years later, I was working with a colleague from our physics department. He was interested in fast pulsed Raman scattering from semiconductors. In this process, two laser pulses are used, with the second one delayed a small amount (typically on the order of a few picoseconds) from the first. In both pulses, the photon energy is larger than the bandgap of the semiconductor. The first pulse was used to create a high density of electrons and holes in the semiconductor. Since the photon energy was much larger than the bandgap, the electrons would be created well up into the conduction band, and could transfer to the satellite valleys for a short period of time (after the pulse). Then, they would move back to the central valley and emit a cascade of optical phonons as they relaxed to the bottom of the band. Now, the role of the second pulse was to measure the optical phonon population. In Raman scattering, the photon creates an electron–hole pair, which then recombines quickly (sometimes). The electron, for example, can emit or absorb an optical phonon prior to this recombination, so that the photon that comes out from the recombination process lies at a different energy than the incoming photon. This shift in energy is the Raman effect. By measuring the intensity of the two shifted energies, one above the original energy of the photon and one below it, you can estimate the occupation number of the optical phonon states—e.g., how many phonons exist in that state. In equilibrium, the number is small, but with the phonon cascade mentioned above, this number can be driven to a relatively large value. By delaying the second pulse relative to the first, one can measure this population as a function of time. This allows one to measure not only the phonon lifetime (the time before it decays to a thermal value) but also how long the electrons stay in the upper valleys. To calibrate this, one needs to have a theoretical evaluation, which my group carried out using computational approaches to the experimental problem. What we found was that in GaAs the electrons remained in the satellite upper valleys only for a small time of order 1–2 picoseconds. On the other hand, in the narrow-bandgap materials like InSb and InAs, they remained in the upper valleys for a very long time, perhaps as much as 10 picoseconds. The reason lay in the physics of the central valley, where the very small value of the effective mass made it relatively hard to transfer back to the central valley.

More recently, we have been working with groups who are interested in making high-electron mobility transistors (HEMTs) with InAs-based materials. The allure of these devices is that they have tremendous potential for working in the terahertz regime, where there are very few options for electronic or optical devices. Our role has been to simulate the devices with the full band structure as well as with the computational simulation of the transport within the devices. As the experiments were pushed forward with actual devices, it was found that these devices could withstand unusually high voltages. As we discussed above, InSb and also InAs would be expected to break down at relatively low voltages due to their small bandgaps. With the InAs-based material strained, its bandgap would be increased to about 0.6 eV. Hence, it was expected that it would show breakdown effects at about this same voltage. Yet, the devices were quite stable at voltages as high as 2.5 V, an incredibly unexpected result. From our simulations, however, we discovered an old friend raising its head. In these devices, there was a very high electric field in the region around the gate, which is normal in transistors. But the electrons would cross this region very quickly, and the electric field was sufficient to transfer them into the upper valleys of the conduction band. More importantly, however, once they arrived in these upper valleys, they stayed there. The time required for them to return to the lower valley was longer than the time they actually spent in the transistor. In the upper valleys, the breakdown voltage was much higher than when the electrons were in the lower valley, and this explained the experimental observations. Physics that we had begun to study almost 50 years earlier was now very important for modern HEMTs pushing into the terahertz region.

Of Lines and Dots

The above thread lasted through most of my career, with bits and pieces coming and going with time. Another thread, which has been almost as continuous, started when I returned to academia in the late '70s. I had left the university life in the early '70s so that I could experience government life. For this, I joined the Office of Naval Research. A few years later, I had the opportunity to move to Colorado State University as the department head for electrical engineering, an opportunity I took as I missed the interaction with young people. A research thread that began at that time was associated with superlattices and quantum effects in semiconductor devices.

In 1976, Douglas Hofstadter published a paper on the energy levels and wave functions in a quasi-two-dimensional system in the presence of a magnetic field. He wasn't the first to do this, as earlier work by the mathematician Harper and the Russian physicist Azbel had shown a number of the effects. But Hofstadter carefully exhibited the fractal nature of the energy levels in the presence of the magnetic field. Normally, the lattice constants in materials are far too small to allow much of this behavior to be observed. But, in the presence of a superlattice, one might be able to see some of the structure discussed by these authors.

To understand the idea of a superlattice, let us return to our idea of the USGS layout of a western city. In this scenario, each block forms what we could call a unit cell, with a lattice constant of one furlong (the edge of the block). In a real semiconductor lattice, the size of the unit cell is about half a nanometer. Before the city has been built up, each block looked just like any other block, and you could create the map just by a translation of one block in the east–west and north–south directions. These would form what we call a two-dimensional lattice (in the parlance of science fiction, this has been called flatland). But the mile streets also form a lattice whose unit cell is a square mile area, which is called a section. Thus, a section contains 64 blocks and 640 acres. From high altitude, one cannot discern the blocks, and the streets look like the lattice. The lattice of the mile streets can then be thought of as a superlattice. If we now put in expressways every 10 miles, then a second superlattice can be created in which the unit cell is 100 square miles in area. Thus, many superlattices can be conceived within our simple layout of the western city.

To create a two-dimensional electron gas in a semiconductor, in correspondence with the two-dimensional map of our city above, we work with heterojunctions. A heterojunction is the positioning of one semiconductor adjacent to a different semiconductor. For example, if we grow AlAs on top of GaAs, we have a heterojunction at the interface between the two materials. More importantly, one can form a two-dimensional electron gas at the interface where these two materials come together, and this can be a very high quality electron gas. Herbert Kroemer and Zhores Alferov shared the Nobel prize in 2000 with Jack Kilby. We tell Jack's story later, but the first two were rewarded for their development of the concept of heterojunctions and their application in semiconductor lasers.

The ideas of the fractal properties can be illustrated by returning to our blocks and city layout. Suppose we continue this layout for a very large city. However, we don't want to go on forever, yet we would like to preserve the regularity with a mathematical trick. We want to take our city and fold it back on itself to introduce an infinite periodicity of the block. The most common method is to just lay our map onto a sphere such as the earth. The problem with this is that the squares become triangles as one approaches the poles, and this breaks the periodicities. That is, folding onto a sphere causes us to have to change the topology of the unit cell. Rather than choosing a sphere, we proceed in a different manner, which will preserve the topology. First, let us take the north–south direction and fold it around onto itself to create a tube, much like the tube in a roll of paper towels or the foam "noodle" toy used in the swimming pool. Now, we can fold the east–west direction back onto itself to make a large donut. If we draw a line on the map in some direction other than the north–south or east direction, then it will wrap around the donut, and probably pass through the donut hole. Thus, we wind a number of times around the outer diameter and a number of times passing through the donut hole. These two numbers give us the so-called winding numbers of the line on the donut. Importantly, this line will eventually wind up closing on itself. In our semiconductor landscape, however, this behavior will change if a magnetic field is applied normal to the two-dimensional electron system. The magnetic field will cause the electrons to curve around the magnetic field in what is called cyclotron motion. With this curvature in the line drawn above, the line will not always close upon itself on the donut. For the line to close, the two winding numbers found above must be rationally related. That is, the ratio of the two numbers must be a rational number. Take for example, 4 and 8, then the ratio is either 2 or 0.5, both rational numbers. But if we take 7 and 11, the ratio (with either as the numerator) is not a rational number, but an irrational repeating number. Only when the two numbers give a rational number, can the quantum mechanics be solved to give an energy level.

What Hofstadter had shown, was that the energy levels were neither continuously available, nor separated by convenient values. Instead, they had a very inhomogeneous distribution, which led to their being called fractal. This fractal behavior arises from the irrational numbers mentioned above. Moreover, it was clear that the energy structure was repeatable with a certain period in both energy and in magnetic field. The question of interest then was could these periodicities be observed experimentally, and also could the fractal behavior be observed. In the GaAs-based heterojunction, the electrons lay just into the GaAs material, so that the basic cell size was the lattice constant of GaAs, about half a nanometer. Certainly, no one was going to see the interesting structures with such a small cell size. Our plan was to create a superlattice with a much larger cell size in order to search for experimental confirmation of the theory.

The idea was to create a surface superlattice which would impose an additional periodicity onto the electron gas, much like the mile streets or the freeways of our city map discussed above. Here, we would use electron beam lithography (a form of photography in which the very small electrons replaced the light of normal photography) to create a set of gates with which to apply an additional voltage to the crystal. This voltage would add to the atomic potential to induce formation of the superlattice. Instead of just 64 or 100 times larger, however, we wanted to make the supercell size some 100,000 times larger than the area of the semiconductor unit cell, or about 160 nanometer on a side. In work beginning at Colorado State University and continuing after my move to Arizona State University, we ultimately made such superlattices and were able to demonstrate the periodicities of the energy structure in both the energy itself and in the magnetic field. The latter was perhaps more interesting, in that we demonstrated that one could see significant transport effects that were periodic in magnetic field, with a period related to the size of the supercell. To us, this was a marvelous verification of an elegant theoretical calculation. But alas, not much technology has come from it, other than a few patents.

In the above discussion, we drew a line in our two-dimensional layout of blocks and used this to define the winding numbers on the donut. To get this line, we ignored the edges between blocks and just passed through them with the line unimpeded. But, quantum mechanically, the line may also be reflected at the edge of the block or cell. If it is reflected from all four sides of the cell, it will still close upon itself (in the absence of the magnetic field) and create a small rectangle that is enclosed within the cell. This rectangle touches the edges at four or more points, the latter if it is not a simple rectangle but reflects many times before closing. This means that we don't have to study the entire map, or the entire donut surface. We only have to study the details of a single cell or block. This is the principle property of the unit cell in crystalline materials. For example, diamond and silicon both have exactly the same crystal structure. Of course, diamond is composed of carbon atoms while silicon is composed of silicon atoms. They have very different properties. Diamond is transparent in visible light and is an insulator (very wide bandgap) and is a very hard material. Silicon is not transparent and appears almost grey in visible light. In addition, it has a small bandgap so that it is a semiconductor, and it is not nearly as hard as diamond. Yet, they have exactly the same structure of their unit cell. There is only a small difference in the size of the unit cell and a small difference in the atomic potentials, yet these differences account entirely for the difference in the properties discussed above. Small differences in the details of the unit cell produce dramatically different

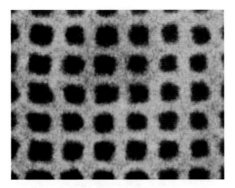

A grid gate composed of gold (light areas) deposited upon GaAs. When a voltage is applied to this gate, a superlattice potential is imposed upon the electrons.

properties in the materials. Hence, studying the unit cell allows us to talk about the properties of the much larger crystal of the material.

In our superlattice created on the GaAs heterojunction, one (super) cell can be isolated as well. The properties of this surface unit cell tells us about the properties of the entire surface superlattice. For the present case, the unit cell is a rectangle (or square) cell, but it could be in many shapes experimentally. Such a cell is called a quantum dot, because it is small on the surface and its properties are controlled by quantum mechanics. Thus, in the evolution of this thread, we switched from studying the large surface superlattice to studies which concentrated on the quantum dot. One type of dot was created by using metallic gates on the surface, where a negative voltage applied to these gates would remove the electrons from the two-dimensional layer at the interface. Thus, the electrons only resided where there were no gates. Importantly, we could provide connections from the large two-dimensional areas to the dot through small openings in the potential known as quantum point contacts. It was after the discovery of conductance quantization in such quantum point contacts by groups in Delft (with Philips Research Labs)

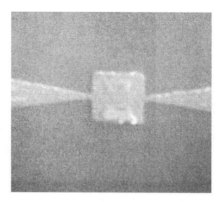

A GaAs quantum dot which has been defined by etching away undesired regions. The dot (light area at center) is about 0.5 micron in size. Current is fed in by the tapered quantum wires through the quantum point contacts adjacent to the dot.

and Cambridge that we (and many others) realized we could access the quantum dots without eliminating the quantum mechanical behavior. I should say that we got seriously into this working with Jon Bird, first when he was at RIKEN in Japan and then as he passed through ASU on his way to Buffalo. A second type of dot, shown in the figure was formed by etching the semiconductor region where the electrons existed. This produced a firm, lithographically defined dot, as shown in the accompanying figure.

When we measured the properties of the quantum dots at low temperatures, we observed that the conductance would show amplitude variations as the Fermi energy in the contacts or as the applied magnetic field were changed. These conductance variations (I will call them fluctuations for want of a better word) were almost periodic in either the Fermi energy or with the magnetic field. Fourier transforming them (in magnetic field for example) showed a clear single frequency peak and occasionally some weak sidebands. These apparently were first observed at MIT, where they also found just a few frequencies. Then, they were studied at a couple of other places where they were interpreted as random fluctuations induced by disorder within the semiconductor. But these were not random fluctuations as the latter would produce a very broad Fourier spectrum, not the single peak which we observed. On the theory side, we were able to show that the conductance would show the equivalent of an energy spectrum that varied with magnetic field, not as dense in terms of energy levels as that found by Hofstadter, but with some similarities. Simulations based on the theory clearly showed the existence of a diamond shaped "scar." Scars are normally thought of as high amplitude areas of the quantum wave function. (The local electron density is proportional to the squared magnitude of the wave functions, so the scar represented regions of higher electron density.) These occur along remnants of classical trajectories, so this means that the trapped classical paths had a similar diamond shape. But scars are normally unstable while we found them to be stable, a result that apparently was related to interactions between different energy states. The interaction had to be dominated by the principle state creating the strong single peak in the Fourier spectrum.

Just over a decade ago, Zurek from Los Alamos published his decoherence theory for open quantum systems, and it became apparent that our observations of the quantum dots was a perfect example of his theory. When the quantum system is opened to the environment, a fraction of the quantum states within the dot would be washed out by the interaction with the environment. On the other hand, there were a significant number of the states which did not couple to the environment and would survive through what Zurek called the einselection process. These surviving quantum states were the *pointer states*. However, since these pointer states do not couple to the environment, how were they affecting the conductance which is measured between the two environment contacts? Jon Bird, and his colleagues at RIKEN had shown years before that magnetic states passing through the QPCs could tunnel to trapped states within the dot and this would appear as either an increase or a decrease of the conductance. The increase would occur with forward tunneling while the decrease would arise from back-scattered tunneling events. We had already shown that the individual amplitude of the fluctuation had the line shape appropriate for a tunneling process. So it appeared that the conductance fluctuations were the result of a process in which the so-called straight-through trajectories, which arose from the washed out states, coupled to the trapped pointer states via a resonant tunneling process. It all fit with Zurek's theory. But could we experimentally "see" the pointer states?

There is an imaging technique that could do the job and which was called scanning gate microscopy. In this approach, you begin with atomic force microscopy. Here, one uses a nanoscale cantilever that flexes due to forces between the tip (of the cantilever) and the surface over which it is raster scanned. By measuring the flexural forces, one has an estimate of the force between the tip and the surface. This process had shown the ability to do atomic scale resolution of surfaces, so that you could "see" (through computer processing of the forces to generate an image of the surface) individual atoms on the surface. In scanning gate microscopy, we modified one of the tips by evaporating metal onto the surface so that we could apply a voltage to the tip. The principle here is that within the quantum dot, the electron density is inhomogeneous, with the density being higher where the wave function scar would be. By applying a negative voltage to the tip, this would induce a reduction in the electron density just below the tip, and this reduction could be measured as a change in the conductance through the quantum dot. This would certainly not have atomic resolution any longer, but a number of experiments suggested that we still had about 5 nm spatial resolution. This was certainly good enough to expect to see the scars in the conductance maps.

This all sounds quite simple, but nothing is simple about doing experiments at very low temperatures, particularly when you are varying a number of different voltages, magnetic fields, and asking a scanning system to work well at these low temperatures. Nevertheless, the system worked sufficiently well, and we could examine a number of different nanoscale systems. For the quantum dots, we finally managed a few well-behaved samples in which we could image the density. But these images did not often compare well with the simulated diamond shaped scar. We have to remember, however, that the pointer states survive because they don't couple to the environment and here we were bringing the environment (the tip voltage) right into the interior of the quantum dot. It was only after we realized that we should simulate the quantum dot with the raster tip in place that real progress was made. Now, the observed conductance map did begin to appear much like the simulation. In the end, we were able to map the pointer states and clearly show that they were not imaginary concepts. In essence, these became the smoking gun for the correctness of Zurek's theory on decoherence in quantum mechanics.

We now had an experimental platform which agreed with a nice theory that allowed us to study the transition from classical mechanics to quantum mechanics, and vice versa. Here was evidence on how to begin with a quantum view of the world and then make the scale change until we arrived at a classical view of the world. This became very satisfying, at least to us.

Unintended Consequences

It is relatively hard to explain what good engineering design really means, because it is relatively easy to screw it up or to succumb to unintended events that ruin the conceptual idea. Generally, the idea is that good design should be fail-safe, which means that if one of the parameters or assumptions fails, the system does not fail and continues to provide the desired results. On the other hand, it is relatively easy to recognize poor engineering design, because you can often see the oncoming disaster as if it were a runaway train. Unfortunately, I have had the opportunity to experience the unintended consequences that can arise when design is not perfect. And, sometimes this creates a real disaster while other times it just leads to inconvenient results.

When I was a young faculty member, our department had adopted the project laboratories. A student, or a small group of students, was assigned a laboratory project, where they had to design the system, put it together and then test it. To accomplish this, they checked out the power supplies and test equipment from the stockroom. I was rummaging around in the stock room one day and came across an unusual piece of equipment. The equipment itself was not unusual. It was a variable transformer, which would deliver between 0 and 120 V from the mains as a control knob was turned, with a 15 V voltmeter attached across it. It was this connection that made it very unusual, and it probably explained why the voltmeter was not working, since the indicator needle was bent and wouldn't move. But the way the two were connected seemed to indicate to me that someone had come up with an incredibly poor design concept. I asked one of the more senior faculty about the items, and after hemming and hawing a considerable amount he told me the story. Now, this may have been just an urban legend, or an old professor's story, but I had seen the equipment and therefore it became a creditable tale.

It seems that the agricultural college was interested in getting into the growing field of artificial insemination for breeding a better grade of cattle. One reason for this was the relatively slow process of hoping that the bulls would be attracted to the breeding process. To push the project forward, they also wanted to use the new technique of electrical stimulation, where an electrical signal was applied to the bull's vital parts and this would lead to the ability to harvest the bull's vital bodily fluids. These could then be used in the artificial insemination project. But they needed the electrical signal to make the entire thing work, so they turned to the electrical engineering department on the theory that the latter would know how to make the necessary equipment in a reliable fashion. Apparently, this was a poor assumption. What they got was the variable transformer with a low voltage meter attached to the output, and a stop glued on the device, so as to keep the knob from being turned too far. This would make sure that only a low voltage was applied. I am sure that everybody can see where this is going, but of course Murphy's law raised its head and the stop became unglued, thereby falling off and being lost. This subsequently led to the voltmeter being destroyed by too much voltage, so that now there was no way to know when too much voltage was being used. Well, time passed, and someone decided to give the system another try, and apparently the corporate memory of the various failures of some parts had disappeared. So, the bull was connected to the equipment and the technician gave the knob some vigorous turns, trying to get the voltmeter to read. This of course provided enormous stimulation to the bull, who it is said performed in an incredible manner, going out in a blaze of glory. Unfortunately, he had to be put down. This loss of a valuable breeding bull naturally created a certain level of anger in the agriculture college, and the equipment was sent back to the department with some obviously unpleasant comments. So, the equipment went back to the stockroom to await rediscovery by a young faculty member.

The idiocy of the entire process lay in the fact that a few pennies spent on an isolation transformer would have prevented the ultimate tragedy. This was in the days when vacuum tubes still dominated much of electronics, and these always required a step-down transformer to provide the 12.6 V necessary to heat the tube cathodes. Such a transformer would have been the perfect choice to add to the system, as these transformers were dirt cheap due to their abundance. Moreover, it would have gone a long way to making the project fail-safe by limiting the maximum voltage that could have been applied. But quick and dirty design often produces tragic results. The world is full of stories of million (or these days billion) dollar aircraft and race cars in which failure is caused by the use of a cheap part rather than a more reliable and more expensive part. But this substitution can also lead to the unintended result.

In those years, I also worked with our systems people in the areas of linear and nonlinear analysis. Now, this was not semiconductors. But I had learned the gospel of the experimental lab (to which I return later) and had lots of time to take up theoretical studies and working with our systems group on some interesting nonlinear problems. The systems group worked closely with air force scientists in New Mexico at one of the airbases. It was during this period that the air force scientists came to us with a problem. We were told that the F4 aircraft had a problem with vibration, and that apparently the inertial navigation system had to be re-initialized after take-off or after any acrobatic flight. This was something that was not supposed to happen. Now, I have never talked with any F4 pilot, so I don't know how true the story was, but this was the situation that was presented to us. I have talked with FA18 pilots during later years when members of the Defense Sciences Research Council visited various military facilities. One of my colleagues asked the pilot how we could improve their electronics to help them in the mission. This is just another example of a poorly framed question, since the pilot is trained to execute his mission with the equipment on hand. Hence, the answer came back that he had everything he needed to do his mission. So, talking to the F4 pilot probably would not have revealed any problems that hindered his mission. Nevertheless, the air force scientists replaced the navigation system on one F4 with an equivalent box that was stuffed with various accelerometers and data recording equipment. We were given the data on massive reels of magnetic tape and also given details of the various flight configurations that generated the data.

Our approach was very interdisciplinary in nature, as we first derived a high order mechanical model of the aircraft and its vibrational modes. We could determine these modes by knowing the mass distribution throughout the airframe. What we didn't know was the various amplitudes of each of the modes, so those were taken as parameters that could be adaptively adjusted to fit various data sets. The input to the aircraft during flight is turbulence that always exists in the environment. This level of this turbulence varied with the speed of the aircraft as well as with the flight maneuvers that were being pursued. Hence, under any flight scenario, the amplitudes of the various aircraft modes would be excited to a certain level. This level would be different for each scenario. These amplitudes were then adjusted for each flight configuration for which we had data. We could use the amplitude of the turbulence as another adaptive variable, but use was made of the known spectrum of air turbulence as determined by von Karmen. Knowing the output data for each segment of flight, we could now adjust the adaptive parameters on our input and aircraft model to yield the observed output data that had been recorded.

It was quite likely that the navigation system had been designed by one group, while a different group (perhaps with a different company) had designed the aircraft itself. This certainly would not have been unusual. The specifications for the navigation system apparently were not detailed enough to prevent an unintended consequence from occurring. What we found in our analysis was that one of the modes of the aircraft had a frequency very close to that of the rubber shock absorbers with which the navigation system was mounted. As a result, the excitation of that vibrational mode would be resonantly transmitted right through the shock mounts and into the navigational system undamped. This was a sufficiently large vibration that it would upset the navigation system in a number of maneuvers, just as was being observed. The fix was simple; just replace the shock mounts with ones that possessed a different resonance frequency. Mechanical systems, just like electrical systems, always have such resonances, and the job of the designer is to keep them away from the desired performance frequencies of the systems. When this is not done, then one gets unintended consequences, and an otherwise excellent design is seriously compromised. Now, I don't know if this fix was ever implemented, but we had given the Air Force a cause and a solution, so I suppose that was the end of our participation.

Sometimes the unintended consequences work in a positive way. Following my graduate work, my wife and I lived in Vienna for a period. Our apartment was on the fourth floor of a building with no elevator. Worse, the heat was a traditional tile stove that was fueled with wood. The wood was stored in the basement and thus had to be carried up all those stairs. But, in usual European design, the apartment had these double windows, with the inner set being spaced from the outer set by a particular uniform distance. I suppose that an architect or designer had worked out that this distance was the optimum to help isolate the outside temperature from that on the inside of the apartment. Another problem was that the apartment had a very small refrigerator. It would hold a few vegetables and some meat and ice, but not much more. One of our discoveries was the market hall a few blocks away, in which there were independent stalls selling a variety of food stuffs. One of our favorites was the wine merchant who sold 2 liter bottles of very good local wines. As you may imagine, there was no way that the white wine was going to go into the refrigerator. However, we discovered that the spacing between the inner windows and the outer windows of the apartment was just the right size to hold these 2 liter bottles. Moreover, it kept them at just the right temperature. I have always wondered whether the bottle designer achieved this magic or if the building designer had taken this into consideration. Or, perhaps it was just an unintended consequence. But the result made our life a lot more enjoyable.

Different Threads

There are different threads that weave through a person's life beyond the technically detailed ones. Two of these in my life involve conferences. In 1980, the Austrian Physical Society supported a winter school on new concepts in condensed matter physics. The winter school was organized by three friends with whom I worked during my post-doctoral time in Vienna: Helmut Heinrich, Günter Bauer, and Freidl Kuchar (in later years, a fourth, Wolfgang Jantsch, took on a good bit of the organizational detail, especially after Helmut retired). This 1980 winter school was famous, not the least because it was the site for the first discussion of the quantum Hall effect, months before the famous PRL was submitted. This first meeting was held in the small town of Mariapfarr. With the success of the first meeting, the winter school became a biennial occurrence and moved a few miles to the west to the slightly larger town of Mauterndorf, sited just below the Tauern pass on the old Roman road over the alps. All of the conferences since have been held in the castle, except for one year when the castle was under repair.

It was perhaps the third of these conferences when I became a regular attendee, usually as a speaker. These conferences were also famous for their format as well as for their scientific content. Much like a Gordon conference, there were sessions

Left: The castle at Mauterndorf. *Right*: The author dressed to direct the awards ceremony following the ski race.

in the morning and in the evening, with the afternoons free for individual activities. As you may guess for a conference site in the middle of the Austrian alps, skiing was a common individual activity. In fact, the organizers would work with the local ski organization to facilitate a race on Friday afternoons, with all attendees encouraged to try their hand at the race. Of course, with attendees from Germany and Switzerland also present, the race quickly became something beyond just a simple afternoon excursion. There were categories for both younger men and older men, as well as for women and teams. Each team had to include an older man, but this didn't lessen the spirit of competition. Now, one year they decided to give a special medal for the team that had the best name. Mark Reed and I thought about this and came up with the winning name—"the last chance saloon ski team." It's the only medal I have ever won in sports. But the winter schools remain well embedded in my history, because of the good science that was discussed here, and the several articles and books that had significant parts written while in Mauterndorf.

In recent years, I served another role at the winter school. Following the Friday ski race, one was required to have a medal ceremony. This took place following the gala banquet on Friday evening, and I was tasked to run the award' ceremony and giving out the medals. Usually, I was dressed in medieval garb that was kept at the castle, and this meant that the event had to be one of fun and enjoyment. So, I took on this task and enjoyed it thoroughly. Alas, all of the organizers mentioned above have retired, as have all of the important people in Mauterndorf who worked with them to make the winter school so successful. Whether or not it continues is an open question, but I suspect my connection with it has run its course.

In the early 1990s, I began my treks to Hawaii in December. This began with a conference called the Advanced Heterostructure Workshop, that was organized by Herb Goronkin (of Motorola at the time) together with a Japanese counterpart. This workshop is held biennially, and has become almost permanently ensconced at the Hapuna Beach Resort on the big island of Hawaii. At about the same time, Chihiro Hamaguchi organized the conference on New Phenomena in Mesoscopic Systems, but held every three years at various locations throughout Hawaii. Now, these two timings leave a empty December every fourth year, so in 1993, we organized the first conference on surfaces and interfaces of mesoscopic devices, with the plan to hold it every four years when that year had a vacancy in December. Finally, in 2007, the last two were combined in the biennial International Symposium on Advanced Nanodevices and Nanotechnology. This secured the fact that there was a conference worth attending every year in December in Hawaii.

These conferences have always been jointly planned with Japanese colleagues. The last one mentioned above is usually funded by a Japanese science program, so they are truly international in character. They also draw a significant attendance from the Europeans as well. So, the science content has been quite high throughout the series. You may have noticed that these conferences have made it clear that there is a major international conference in Hawaii every December, and this conference is associated with the field in which I work. There are a great many colleagues who join me annually at these conferences, so perhaps they share the thought that underlies this trend: A year that goes by without a week in Hawaii in December is just a wasted year!

Chapter 3

Rise of the Chip

Earlier we mentioned Jack Kilby and the invention of the integrated circuit in 1958. At the time, Jack was a relatively new employee of Texas Instruments in Dallas, Texas. The company had a policy that everyone would take their vacation at the same time, so that they could just shut down the manufacturing facilities. The problem was that Jack was such a recent hire, he didn't have enough accrued time to take such a vacation. So, he had to remain in the facility finding something with which to keep his mind busy. Left to his own devices during this period, he realized that an entire circuit could be manufactured more efficiently if it all lay on one integrated piece of material. Over the period, he began his work to integrate a complete oscillator circuit on a piece of germanium, one of the early materials in the semiconductor industry. When the summer was over, he arranged a demonstration for his managers, and showed how this small piece of germanium made a modern oscillator. The patent was quickly filed. As they say, the rest is history.

I had the good fortune to meet Jack when I was at the Office of Naval Research. While we sponsored basic research, mostly at universities, some folks in the pentagon thought that enough

50 Years in the Semiconductor Underground
David K. Ferry
Copyright © 2015 Pan Stanford Publishing Pte. Ltd.
ISBN 978-981-4613-34-7 (Hardcover), 978-981-4613-35-4 (eBook)
www.panstanford.com

of our programs edged upon industrial developments that we should get some coordination on that score. The pentagon had an advisory group, the Advisory Group on Electron Devices (AGED), which coordinated development programs among the various services. So, it was agreed that we would "coordinate" our semiconductor research programs with AGED as a way to keep them informed of research advances, while they could politely tell us when industry was well ahead of our ideas. Jack Kilby was a member of this group, and this was where I had the chance to meet him. While I left ONR shortly after this, I had the chance to continue to interact with him over the years.

As is common in science, a novel idea often happens at several places. Perhaps it is that the idea has a set of predecessors which come together as a logical conclusion to several people at about the same time, since the prior knowledge is available to all. In this case, Robert Noyce at Fairchild had a similar idea of integrating a circuit into a single piece of semiconductor. Noyce had come to the west to join Shockley Semiconductors, but left with other disillusioned engineers to join Fairchild, where they would pursue his ideas for integrated circuits. His patent was slightly after that of Kilby, but it went somewhat further. In 1968, he left Fairchild with Gordon Moore to found Intel. It was here that the microprocessor was invented, and thus started a major intellectual development that has affected us all. It is the microprocessor and Moore's law that engendered the information age that has blossomed in the past four plus decades.

Moore's Law

Gordon Moore was educated at Berkeley and Caltech, after which he joined Shockley Semiconductors. And he was one of the disillusioned group who left with Noyce to join the new effort at Fairchild Semiconductors. His original formulation of what became known as Moore's law was an empirical observation in 1965 that the number of transistors that they had been able to put onto a single microchip had increased by a factor of 2 every 18 months. You have to understand that this was only a few years after the first integrated circuit, so he could have only seen perhaps 3 doublings. Yet, he was visionary enough to understand that this was an important statement about technology and the growth of the manufacturing that was possible in the semiconductor field. While there is a scientific underpinning, which we will get to a little later, the driving force for Moore's law is really economics. The transistors in an integrated circuit are laid out in a planar fashion, much like the streets in a USGS survey for a town. The transistors are placed like our blocks and houses. Hence, if I reduce the size of the edge of a block (be it a transistor or a city block) by a factor of 2 in each direction, the area is reduced by a factor of 4. Moore observed that this occurred every 3 years, so that the factor of 2 occurred in half this time, or 18 months.

In Moore's original view, the law was driven by three primary factors: (1) reducing the transistor size and therefore the semiconductor real estate upon which it was situated, (2) increasing the size of the small integrated chip itself, and (3) circuit cleverness by which the number of transistors needed to create a function could be reduced over time; hence the number of functional units that appeared on a chip could be increased. This last factor meant increasing the computing power on the chip as each new generation appeared. Because the basic cost of manufacturing the chip has not dramatically changed over the past five decades, the cost per functional unit goes down as more such units appear on each chip. This leads to exponentially increasing computing power for a relatively constant price. Thus, the economic driver is the cost of silicon real estate, and this drives Moore's law.

Since the end of the twentieth century, however, the size of each microchip has remained fairly constant, a result of the need to get the energy dissipated out of the chip. Generally, one lowers the voltage and current proportionally to the size decrease, so that the power per unit area remains constant. However, with today's complementary MOS design, the power dissipation occurs mainly when the transistors are switched. So raising the clock frequency results in more energy dissipation in each block, so one tends not to raise the clock speed for this reason. As a result of holding the chip size and speed constant, factors 1 and 3 above have become more important. For example, Intel's great leap forward at the end of 2012 was the tri-gate transistor. Here, the transistor is no longer planar, but is largely vertical, placed on the sides and top of a vertical fin (like a shark's fin, sticking up out of the body). With this new topology, shrinking critical length (the gate length) to 22 nm could be achieved while still using less horizontal silicon real estate. Nevertheless, this 22 nm is only about 100 times the spacing between individual silicon atoms. Clearly, this evolution cannot continue down this path much longer. The end of Moore's law has been predicted for more than two decades, but it continues to move forward.* While the tri-gate has been one such revolution, the limit to gate length implied above puts more pressure upon the third factor to continue to grow the power of integrated circuits.

Death of a Research Track

As technology progresses, new things come along. But, often, many approaches in both research and life become obsolete. Unfortunately, the growth of the microchip according to Moore's law destroyed one of my continuing research programs.

Along with creating the surface superlattices discussed in the last chapter, we used electron beam lithography to make very small transistors. Usually, the minimum feature size is the so-called gate length, which is the width of the gate metal line. The gate itself is biased with voltage to control the flow of electrons between the input (source) and output (drain) terminals. Our goal was to seek to make really small devices, whose gate length

---

*In fact, Intel has made an amusing video poking fun at Moore's law detractors: http://www.youtube.com/watch?v=D3dKbq5AXz8

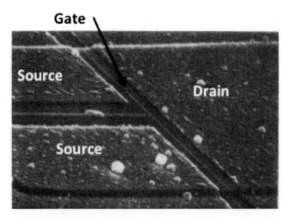

A photomicrograph of a high-electron mobility transistor made at ASU. Gate lengths as short as 25 nm were fabricated.

was in the 25–30 nm range. This would make them the smallest transistors in the world at that time. While we succeeded, there was intense competition from the French and the Japanese, so it was difficult to tell whose devices were smaller on any given day.

We succeeded in measuring the microwave performance of these transistors, and found that we could achieve a cutoff frequency as high as 167 GHz. The cutoff frequency is that frequency at which the gain of the transistor is reduced to unity. We know now that these devices suffered from source and drain contact resistances that were far too high. These series resistances dramatically lower the gain. Nevertheless, some two decades later, the cutoff frequency for high electron mobility transistors has only been raised by about a factor of 5 through strong industrial efforts. This has been achieved by moving to strained InAs for the active channel in the transistor, a material we discussed in the last chapter.

But this line of research is now gone, as it has become irrelevant for university work. The truth is that Intel now produces many thousands more such small devices every day than the population of the earth. Their current design has a gate length of 22 nm, with smaller ones due in 2014. Hence, no matter how successful we were, the industry has put us out of business with their progress. Of course, there are a few university groups who still push to improve microwave transistors through new materials and clever device design. Our thrust, however, was the miniaturization, and this is where Intel, and other chip makers, have excelled.

Scaling Theory

As mentioned above, there is a scientific underpinning to Moore's law, and that is the scaling theory. Robert Dennard was a scientist at IBM for his entire career. In 1968, while at IBM, he invented the dynamic random access memory circuit. As Dennard tells it, when he told his boss about this, he was advised to take two aspirin and come back tomorrow. Nevertheless, this memory technology was revolutionary, and is standard in every computer built today. He was awarded the Tokyo Prize in 2013 for this technological breakthrough. But it is his other big result in which we are interested. By 1974, he was aware that the metal-oxide-semiconductor (MOS) process was the route to the future (there was a controversy brewing over whether continued use of the bipolar transistor or a change to the newer MOS technology was the wave to be followed for the future). He and his colleagues at IBM published their scaling theory for the MOS transistor. We have already mentioned the important factors of reducing the size of the block, in this case the transistor dimensions, to reduce the area. But in order to properly scale all factors of the transistor, all voltages must be reduced by the same factor. If this is done, the current also goes down by this factor, and all of the electrostatics within the transistor are preserved. That is, the internal electric fields all remain as they were in the larger device. This means that the transistor will perform just as the previous, larger one performed. So, the scaling theory provided an approach by which the performance and function of the transistor was exactly preserved by proper scaling principles. If a dimension was reduced by a factor of 2, the area of the transistor went down by a factor of 4, just as observed in Moore's law, and this would cover two scaling periods. Moreover, the power per unit area of silicon would not change, because the voltage and current are each reduced by this factor of 2. Unfortunately, the speed was usually increased so that the energy dissipated per unit area increased. This led to the thermal problem mentioned above. But once you could understand the concepts behind the scaling theory, the circuits could be redesigned with a new driving force—seeking low power dissipation. This has been a mission for some three decades now, and the newer, faster, and more complex chips in the present generation actually dissipate less heat than those of a decade ago.

If we could summarize the tale of Moore's law and the scaling theory, and the work of Jack Kilby and Robert Noyce, it would be that the manufacturing process is exceedingly important. But this also requires that good design principles are followed, and that no unintended consequences raise their ugly heads. Indeed, most modern semiconductor corporations spend more on manufacturing research than they do on basic research for their devices and chips. Today's transistors often include strained regions, different semiconductors such as germanium included with silicon, high dielectric constant oxides, and of course the new tri-gate structure. Each of these novel advances required considerable effort to make them manufacturable with the precise dimensional control required for a cost effective integrated circuit. Today's wafers, in which the chips are made, are 300 mm in diameter, and growth to 450 mm in diameter is on the horizon. Yet, the chip itself is only 1–2 cm^2. So, there are a lot of chips on each wafer, and each chip may contain of the order of 10^{10} transistors (in 2013). Between the silicon body of the transistor and the gate is an oxide. If this oxide is to be say 10 nm thick, then it must be 10 nm, and not 9 nm or 11 nm, across the entire 300 mm size of the wafer. This is an incredible accuracy which must be maintained across the entire wafer, and of course requires unbelievable manufacturing skill and technology. Charles Babbage was mentioned in the last chapter. The major reason he failed to build his seminal computer was that the manufacturing technology required for his effort just did not exist at that time. So, hand in hand with Moore's law for the chip is a parallel advance in manufacturing technology. But this has to go hand in hand with good engineering design.

End of the Mainframe

In the early 1980s, I joined a group named the Materials Research Council. It was subsequently renamed the Defense Sciences Research Council, and is a consulting group to some of the research offices of the Defense Advanced Research Projects Agency (DARPA). We would meet for a few weeks each summer and carry out a number of studies for DARPA. These studies usually examined closely a (narrow) field of research and made prognostications about the future and how DARPA might impact it through various funding actions. Near the end of this particular decade, I was involved in a study directed toward the very high speed integrated circuits that were being pushed by DARPA. As part of the study, we tried to forecast how mainframe computers and microprocessors would evolve over time. Of course, this evolution was determined by Moore's law and the scaling theory that supported it. But there is another factor that is important in projecting the future computing power for both mainframes and microprocessors. This is the number of pins that are required to support a given set of gates. The pins connect data, power, and control signals to the chip. They also allow you to plug the chip into a socket for easy installation and exchange.

Up until the mid-1980s, it had generally been the view that continued increases in packing density for very large scale integration, which was the term used to describe the growing density in the microchip, would require a catastrophic increase in the number of pins. This view resulted from early studies of the chips in mainframe computers, and suggested that the number of pins increased as a power law of the number of gates. The relationship was termed Rent's Rule. From these early studies, it was suggested that the exponent (the power to which the number of gates was raised) was about 2/3. I had undertaken a study of the pin count in microprocessors and found a different result, in which the exponent was only about 0.2–0.25.* This difference is quite significant. If we think about a chip with N gates in a square array (much like our city map used earlier), we can connect pins efficiently to the gates that are along the periphery of the chip. The edge length around the chip scales as the square root of the area of the chip. For this perfectly square chip with pins maximally around the periphery, the number of pins goes as the square root of the number of gates, or leads to an exponent of 0.5 in the above Rent's Rule.

The difference in the exponent is largely due to the type of chip that was involved in mainframe computers at that time. Typically, these are so-called gate arrays. As the name implies, the chip is an array of gates without a defined interconnection plan. This is so that the generic chip can be used in many different applications by different interconnections of the gates. But these different interconnections are implemented external to the chip. Hence, more pins are required to provide for the extra communication for the interconnects. The significance of Rent's Rule lies in the pin count, but just as importantly in the effect the exponent has on the average interconnection length for the chip. This average interconnection length is an average over all possible placements of the gates, and is usually measured in terms of the size of the gate blocks (a furlong in our city blocks), called "cell pitch." Chips, like the gate arrays, that have an exponent larger than one-half are found to have very long interconnections, which severely limit the speed of the chip.

On the other hand, the microprocessors may be defined as a highly functionally partitioned architecture and this should

---

* D. K. Ferry, *IEEE J. Sys. Devices* 1(4), 39 (July 1985).

lead to a much smaller value of the exponent.* Hence, the value of the exponent I had found for the microprocessors was in agreement with these expectations. This has a consequence that may be expressed in terms of the average interconnection length. Since the signal doesn't leave the chip until the end of the computations, one expects that the interconnections are largely from one gate to the next. In fact, using the square "city map" type of circuit layout, one can compute the average interconnection length[†] and this is found to be only about three circuit pitches. The importance of this is that the average interconnection length is independent of the size of the chip, and scales with the other factors on the chip. Thus, this average interconnection length (in terms of this pitch idea) is unchanged as Moore's law proceeds to higher density chips.

It is perhaps more interesting that the partitioning of the chips and the resulting Rent's Rule for the number of pins required can be related to an "information dimension" of the chip. Mandelbrot actually makes the connection by pointing out that this dimension is a fractal dimension, given by the exponent. When the exponent in Rent's Rule is 2/3, as found for the gate arrays, this dimension is 3. That is, one requires a three-dimensional environment for the small two-dimensional package of the gate array. This extra volume requirement is the result of the very long interconnects that are required. On the other hand, the microprocessors result in an information dimension that is only 1.25–1.33. That is, the information flows through the microprocessor in an almost one-dimensional manner. Again, this is the result of the very short average interconnects in the microprocessor and its functional design.

Moore's law gives a connection between a point in time and an integration level (number of transistors on the chip). The scaling theory brings this together with Rent's Rule to define

---

*J. C. McGroddy and M. I. Nathan, *Intern. Electron Dev. Meeting Tech. Dig.*, pp. 2–5 (IEEE Press, December 1982).
[†]W. E. Donath, *IEEE Trans. Circuits Sys.* **26**, 272 (1979).

an architectural limit for gate arrays and microprocessors. With these, one can define a prototype mainframe and a prototype microprocessor and discuss the properties and limits of each. That is what we did in the study for DARPA. We chose an integration level, in terms of the number of gates and/or transistors, that was appropriate for a few years beyond the time of the study. Then, we could give a design in terms of gate arrays which included e.g., 5000 transistors each. This led to a conclusion that we could only package a few dozen arrays onto a multi-chip module that required cooling processes to handle the heat. The long interconnects required significant power in the gates in order to send the signal down these long interconnects. This led to high chip temperatures and the need for extra cooling processes. In fact, this came very close to a mainframe that was being built at the time. The manufacturer was generating publicity about the cooling in their multi-chip modules. My view was that when an electronics company is touting their mechanical engineering, they are probably near the end of their product lifetime.

Similarly, we could use the total number of gates to define a microprocessor and then determine its heat requirements and its maximum operating frequency, keeping everything within the parameters set by Moore's law. This microprocessor dissipated orders of magnitude less heat than the mainframe. While it was right at the limits of how much heat could be removed from the silicon chip, it proved doable as the industry progressed in the subsequent few years. As a result, it was clear that the microprocessor was the preferred approach to implementing the designed computer.

Another thing which came out of the design study was the fact that one could estimate how the computing power would advance with time. It was clear that the mainframe had a much lower increase in computing power with time than the microprocessor. But on the basis of viewgraphs provided to us, it was clear that our predictions did not disagree with our mainframe manufacturer. The astonishing (in retrospect, we should not have been astonished) news was that the progression of Moore's law meant that it would be only a few years until the microprocessor possessed more computing power than the mainframe. This would be the death of the mainframe. So, our recommendations to DARPA that they needed to continue support for the very high speed integrated circuits program was the way in which to drive the technology forward. And, it proved to be the case. Now, I don't think that we discovered any radical new science in this study, but the process certainly served to open our eyes as to the trends of the computer industry.

To illustrate this rapid progress, we note that in 1975, the Cray-I was introduced with a performance of 80 million floating point operations per second (80 MFLOPS). A decade later, their top mainframe was the Cray-2 with a performance of 1.9 billion FLOPS (1.9 GFLOPS). In 2006, the Intel Core-2 Kensfield chip was 20 times more powerful than the Cray-2. A six core Xeon chip has shown a performance of 258 GFLOPS. Today's "mainframe" is a supercomputer using massively parallel processing and contains a few million microprocessors working together. The microprocessor that displaced the old mainframe has become the "gate array" of the new supercomputer. As a result, we live in an "arms race" of computing power. Every six months (June and December), the latest results of the world's fastest computers (the world's most powerful computers as rated by their performance in terms of GFLOPS, or the newest version PFLOPS or 10^{15} FLOPS) are announced.* Of course, with the announcement comes massive amounts of publicity touting the fact that the world's most powerful computer lives at such and such a site. At the time of this writing, that place lies in China.

More Than Moore

While many people have suggested the end of Moore's law was eminent as early as 40 years ago, it has continued onward.

---

*http://top500.org

But with critical lengths pushing below 20 nm, we have to be concerned that the distance between two Si atoms is only about 0.23 nm. The simple fact is that Moore's law cannot continue forever. As a consequence, there is considerable trepidation in the industry as to what will happen when the end is reached.

We remarked above that Moore considered three factors that contributed to the evolutionary growth in the number of transistors on each integrated chip. One of these was the size of the chip itself, and this has been relatively constant since the early 1990s. The limit to the critical size would then take out a second factor—that of decreasing the transistor size. The third factor, circuit cleverness, still remains. This factor deals with the issue of continuing to increase the processing power per unit area of Si. Much can be done here to continue the growth envisioned by Moore. Extending the chip into the third dimension has been envisioned by a great many people as a prospective solution. In a large sense, the move by Intel to the tri-gate vertical transistor is a step in this direction. But much more can be done to extend the chip into the third dimension. Many have suggested layering the chips, much like the floors of a large building. Another idea is to use new materials to create transistors sited within the many layers of metal interconnects lying above the Si transistor plane. Such an approach could lead to a changeable architecture within the chip, so that certain arrays of transistors could be reprogrammed for different functionality as part of the overall computation.

Since the first transistor, logic and computation have been carried out by the movement and storage of charge; each electron moving in a wire carries a bit of charge with it. All transistors today are charge control devices. One question that the industry has asked, and continues to investigate, is whether or not there exists the ability to do logic and computation with a non-charge-based paradigm. One example is to use the electron spin or magnetic flux. The idea for the use of magnetic flux has been around for quite some time, and there was an all magnetic computer in the early 60s. But it lost out to the new integrated circuits. There are some among us who feel maybe it is time to reexamine this concept, perhaps coupling it to novel processing techniques such as cellular automata. There already exist today memory devices which marry the idea of a magnetic tunnel junctions and transistors to make non-volatile memories. But in these tunnel junctions, the local magnetic field of one magnet is switched to turn on or off the flow of charge through the junction and to the transistor. So, this doesn't really get away from the use of charge. Nevertheless, a number of groups are pushing into other ways to use the magnetic flux in logic applications.

Magnetic cellular automata uses a somewhat different approach. Small nanomagnets are fabricated into a logical array. Each such small magnet can be polarized into one of two directions. At this point, these nanomagnets are no more than other proposals for magnetic logic. Instead of the flux itself, it is the direction of the flux which is used for the state of the individual bit. It has been argued that this approach is the most suitable existing technology for building future magnetic logic circuits.[*] The logical function is defined by the physical placement of individual nanomagnets, or small groups of these nanomagnets.[†]

Whether or not the above approach ever succeeds will depend upon many factors, not the least of which is whether or not normal Si chips can continue to exist and grow enough to keep Moore's law alive. The Si microchip industry is an enormous component of the world's gross domestic product. It will not change dramatically the concepts of design and manufacture of the microchip unless it is forced to by a catastrophe. Will a soft ending of Moore's law avoid such a catastrophe? I suspect so.

---

[*]G. Csaba, W. Porod, and A. I. Csurgay, *Int. J. Circ. Theor. App.* **31**, 67 (2003).
[†]A. Orlov, A. Imre, G. Csaba, L. Ji, W. Porod, and G. H. Bernstein, *J. Nanoelectron. Optoelectron.* **3**, 1 (2008).

Chapter 4

Challenging Physics

In the early '60s, Rolf Landauer of the IBM research laboratories was worrying about the physics of the computing process. His goal, of course, was to help IBM find paths to bigger and better computing machines. But in this process, he apparently made a mistake. In his work, he claimed that erasing computer bits was tantamount to physical dissipation. In fact, there is no such connection, as many scientists have shown since the early work.* The issue lies in the concepts of *phase space*. If we have a single physical particle, then classically it has a position in three dimensions and a momentum, or velocity, in three dimensions. Thus, it takes six dimensions to express its position and velocity. If we have N particles, we need $6N$ dimensions to express fully all of the positions and velocities. This is the phase space of the N particles. Now, the issue is that if the particles lose energy, their momenta are reduced, hence it requires less volume in the $6N$ dimensional space to contain the points that describe each particle's position and momentum. We say that this compresses the phase space occupied by these particles, and this compression

---

*See, e.g., J. D. Norton, *Stud. Hist. Phil. Mod. Phys.* **42**, 184 (2011).

---

50 Years in the Semiconductor Underground
David K. Ferry
Copyright © 2015 Pan Stanford Publishing Pte. Ltd.
ISBN 978-981-4613-34-7 (Hardcover), 978-981-4613-35-4 (eBook)
www.panstanford.com

represents a dissipative process. On the other hand, a computer bit has two values, 0 or 1. Hence, it has one dimension which has only 2 points in the dimension. If we have N bits, then we have N dimensions in this phase space. The issue is whether or not there is a connection between these two phase spaces—that of the physical representation of the computer bits and that of the logical representation of the bits. Landauer suggested that there was a connection, and if we erased a bit, that would entail a physical dissipative process which compressed the physical phase space. It is this suggested connection that is deemed to be wrong these days. That is, the erasure of a random logic bit does not require any corresponding reduction in the computer bit phase space, since the states still exist. It may or may not lead to a reduction in the physical phase space.

Landauer also had come to the conclusion that erasing a physical bit of information would require an energy dissipation of $k_B T \ln(2)$, where k_B is Boltzmann's constant and T is the temperature.* This argument was apparently based upon the information theory argument about this dissipation being required to distinguish between two states of an incoming signal. In later work, Bob Bate from Texas Instruments showed that this energy dissipation was required to switch between two energy levels in a quantum system reliably.[†] In essence, Bate was saying that one needed to dissipate this much energy to maintain the electronic device in the non-equilibrium state required to represent the value of the bit. This will be an important point below.

A Thread That Should Die

While the approach of Landauer seems now to be in error, the bigger error came from a colleague of Rolf's, whom we will simply

---

*R. Landauer and R. W. Keyes, *IBM J. Res. Dev.* 5, 183 (1961).
[†]R. Bate, in *VLSI Electronics*, vol. 5, ed. N. G. Einspruch (Academic Press, New York, 1982), pp. 359–386.

call Dr. D (and the choice of D has no relation to any person's initials). Turing, in his famous paper outlining what became the modern computing machine, erased information (the bits) when they no longer were needed. Dr. D came along and said that, if Landauer were correct, then by not erasing any information, we could compute without any physical dissipation. Even if Landauer were correct, there is no basis in physics for asserting that the reverse of his assertion would be correct. We have many such examples in everyday life of such one directional statements. For example, all lagers are beers, but all beers do not have to be lagers; all bourbon is whiskey, but not all whiskeys are bourbon, and so on (in 2014, there was an ongoing legal battle over what could be called "Tennessee Whiskey"). Indeed, it is easy to prove that no such reverse to Landauer's assertion is possible by a simple counter argument. No one would say that the physical transistor structures that are used today do not dissipate energy. Just check the temperature of the microprocessor, and you will observe the truth of this statement. But it is quite easy to design, using these very same transistor structures, a logical gate that does not throw away any information. Thus, it is exceedingly easy to construct a computer chip which preserves all physical information but clearly dissipates energy. Hence, the inversion of Landauer's idea put forward by Dr. D is clearly wrong. But, as John Norton has said, in an email about the topic, "it seems that pointing out fatal flaws in a clear and simple way is just not enough to stop the program."

Dr. D spent considerable time investigating mechanical systems with a goal to find some that would not be subject to friction. Friction is a fact of life that we cannot avoid. In essence, however, what Dr. D was seeking was a perpetual motion machine. There have been known attempts to construct such a machine since the middle ages. What Dr. D desires has come to be known as a perpetual motion machine of the first kind, in which the machine continues to run forever without any input of energy and also deliver work as an output. Here, the work is taken to be the output information that is computed, as work is required to drive the output indicators. Generally, such a machine violates all known laws of thermodynamics and physics. Nevertheless, Dr. D is in good company because a great many patents have been sought for such perpetual motion machines (apparently the patent office wisely now requires a working model before processing the patent application).

My group became involved in this discussion in the early 1980s, somewhat after Dr. D served up his ideas for a dissipation free computer. Two bright post-docs went through his work and came to the conclusion that he had missed (or neglected) several physical laws in his arguments. When we published this information,* we were pummeled by his acolytes. Unfortunately, these acolytes were no better acquainted with the laws of physics than he was.

An important part of any computational process is measurement. In the Turing machine, the measurement is reading the information on the tape of the machine. This tape corresponds to today's dynamic random access memory, or DRAM, where the information is stored. This is not a random process, because the computation must proceed in a manner to find the desired answer. The machine must be forced along the desired logical path. This is the purpose of the computer code—the program—which describes the desired computation. At each step of the process, we read what is on the tape and then apply a rule, given by the program, to move to the next state. Computation therefore consists of a series of steps, which force the system from one state to another, the latter of which is the logical successor state. If the system thermalizes, the process becomes random and only noise is produced. This often leads to the dreaded "blue screen of death" that appears on your computer at the most inopportune times. It is the competition between measurement and the thermalization process that

*W. Porod, R. O. Grondin, D. K. Ferry, and G. Porod, *Phys. Rev. Lett.* **52**, 232 (1984).

requires energy to be dissipated. It is this energy which Bate calculated, and is required to keep the two states of information from thermalizing. One also can say that the presence of noise is always there, so the measurement must be made within this environment, and overcoming this environment may entail even more dissipation. This tells us that there is a contradiction between systems that can be used for computations and systems which are in thermodynamic equilibrium. Only systems with a deterministic time evolution, and that are secured from noise by being maintained in an ordered non-equilibrium state, can be used to represent in a physical manner the deterministic process of computation.

The interesting aspect of this need for the ordered, or steady-state, system in the non-equilibrium state being maintained by dissipation is that this is a phraseology that was introduced by Landauer himself.* In particular, he pointed out that an inflow of energy was necessary to allow these systems to maintain a steady state. Moreover, he also pointed out how such systems were more strongly subject to being disrupted by noise. So, to try to take a single idea from Landauer and jump to a conclusion opposite to Landauer's own contributions to non-equilibrium thermodynamics seems to be rather foolish.

It also turns out that Dr. D should have asked the question as to whether a machine with reversible logic could even do a computation. What is often overlooked is that Turing, in his early work on the computer, actually addressed whether or not one could evaluate a function or to compute a value at the end of the computation. To achieve this end of the computation, the machine had to stop. The logical result was that if the machine didn't stop, the evaluation of the function could not be computed. Reversible logic has some problems in this regard. The logic gate maps an input into an output. If this gate is reversible, then there must be a unique input state, as well as a unique output state. If we take a group of gates, let us say a group of M gates, then the

---

*R. Landauer, *Physics Today* **31**(11), 23 (November 1978).

number of unique states is 2^M for our binary machine. So, a state is described by the setting of each of the gates; the system state is set by the collective setting of the individual gates. When we move from one system state to the next, we can think about the mapping as being represented by a $2^M \times 2^M$ square matrix. Because each state has a unique predecessor and a unique successor, each row and column of this matrix must contain a single 1. But such matrices are well known to be representations for the cyclic group. That is, the system states all live on a periodic ring. Motion from one state to another just takes the machine around the ring in a periodic never-ending fashion. This machine will never stop—there is no stop state that satisfies the reversible requirement. If this machine does not stop, then it cannot compute the value of a function according to Turing.

But, you may say, I am familiar with periodic structures, because we use them all the time in the theory of condensed matter systems. Yes, we use this behavior in computing the band structure and the phonon vibrational spectrum all the time. But, remember, an electron that is free to move throughout the crystal within one of the energy bands has periodic motion, and just oscillates in space (the so-called Bloch oscillation) with no average velocity over time. We need to introduce scattering to allow the electron to gain an average velocity. That is, we need to introduce dissipation to create current flow in our condensed matter system (assuming that it is not superconducting, of course). Current flow is a result of a non-equilibrium state which is stabilized by dissipation. In fact, these realizations have come to the world of quantum computing (which we discuss later) through the use of dissipation to stabilize an entangled steady state.* So, while the roots of quantum computing may have grown from the preaching of Dr. D in his early days, modern efforts in this direction consider the manner in which coherence can be maintained while limiting the level of dissipation. The effort is

---

*See, for example, Y. Lin, J. P. Gaebler, F. Reiter, T. R. Tan, R. Bowler, A. S. Sørensen, D. Leibfried, and D. J. Wineland, http://arxiv.org/abs/1307.4443v1.

to find real world quantum computers and not to demonstrate perpetual motion machines.

But there remain devotees of Dr. D, who assert that there is no need to dissipate in the computing process. In essence, this is a clan devoted to perpetual motion machines, regardless of the real physics involved. It is about this clan, that Norton's statement above refers, as they are the mafia that wants to move this illogical program forward. Perhaps the best way to end this section is with a final remark from Rolf Landauer, who started all this mess. In an email shortly before he died, Rolf admitted that he had never believed in dissipation-free computing. That is, he had never wavered from this view that non-equilibrium systems required dissipation.

Weird Philosophies

Near the end of August 2013, Nasdaq lost its connection to the internet, with the result that stocks in a significant number of companies could no longer be traded for several hours. One financial reporter, R. W. Forsyth of *Barron's*, then posed the question as to whether these companies actually existed if no one could trade their stocks? Now, this is an interesting question, and gets right to the heart of some philosophical outlooks on life. The point is phrased another way by the makers of Tostitos, which is the all-American chip necessary to make a party. On the package is the comment "Double dipping—If no one catches you, it never really happened." Both of these are plays on that age old question, which we all learned as children: "If a tree falls in the forest, and no one is there to hear it, does it make a sound?" Now, I have rather simple answers, at least to the first of these parables. Since I am not a high speed computer trader, I never noticed the Nasdaq outage. Hence, in my universe it obviously never happened, at least within the philosophy expressed by Mr. Forsyth. As for the Tostitos, the comment is even simpler. By the time I finish the first bite, there is not enough left of the chip to warrant a second dip! I have never been accused of eating delicately in the presence of chips and salsa. In fact, when confronted by a plate of nachos, a colleague from the UK has compared me to a whale eating krill.

When the tree falls in the forest, both the motion of the tree through the air and the impact when it hits the ground will create a compressional wave (in the air) that propagates away from the site. Physicists refer to these compressional waves as sound waves, or acoustic waves. Similar sound waves arise from the passage of a plane, a missile, or even a car, as well as from the electric discharge we call lightning. And, of course, we surround ourselves with equivalent electrical sound reproduction systems which allow us to enjoy music. A realist would argue that all of these sounds exist whether or not anyone is there to hear them (or to observe the lightning). But not everyone is a realist. There exist others to whom the sensual experience is paramount. Sound does not exist until it is experienced by the body, mostly by the ears. For these, the world is defined by their senses. Certainly, one can say that the world in which they exist is interpreted by their senses, but is that the only part of the world that exists? You may ask just why anyone would care about these philosophies. But they are important and go right to the heart of modern life. We live in a world surrounded by quantum systems, from the microprocessor discussed above to the laser scanners used in the grocery store to prepare our bill and the baggage handlers who try to assure that our bags arrive at our destination with us. Especially in these last two examples, we would greatly prefer that our bill and our baggage arrival were deterministic (and causal) events. We would be very unhappy to have these events determined by how the clerk and baggage handler felt at that particular moment. We do not want to have the events be determined by random probability.

Ernst Mach was an Austrian physicist and philosopher who did most of his work in the second half of the 19th century. We mostly know him from his work on the effect that different media had on light, particularly when that media is under different forms of stress. In addition, most of us are familiar with his Mach number, which is the ratio of an object's velocity to the speed of sound. But the present issue is with Mach's philosophical view of the world. To Mach, physical theory should be based entirely upon directly observable events. For his entire life, he argued that Boltzmann's study of statistical physics and the random motion of atoms was irrelevant to physics because one could not see the atoms (and so he was of the opinion that there was no such thing as an atom). I suppose that we would label Mach a logical positivist, a form of philosophy that bases all knowledge upon sensory perception. Indeed, it is within the vein of this view that the above parables have their origin. If sound only exists when we sense it, then of course the falling tree makes no sound when we aren't there. The extreme of such a view leads one to limit science to only those aspects of nature that are directly observable, and hence Mach's dislike of atoms. Now, this view of the world might have remained limited to children's riddles were it not for Neils Bohr. It seems quite likely that Bohr was aware of these views and he seems to have imported them completely into his view of quantum mechanics. This view is interpreted such that there are only possibilities, or probabilities, until the measurement is made. That is, in Bohr's world, the measurement defines the quantum system. This seems to be a full reflection of Mach's view of the world. And, there is considerable evidence that much of the debate between Einstein and Bohr hinged on such a philosophical interpretation. Einstein was a realist, and his comment about believing that "the moon is there even if I don't look at it." seems to go directly to a dislike of Bohr's philosophical base for quantum mechanics. Thus, Bohr would have our life devoid of causal, deterministic actions when quantum mechanics becomes involved. In fact, Bohr went so far as saying, "If in order to make observation possible we permit certain interactions with suitable agencies of measurement, not belonging to the system, an unambiguous definition of the state of the system is naturally no longer possible, and there can be no question of causality in the ordinary sense of the word."* This is certainly a Mach-like view, and it is easy to see how Einstein's realism would lead to many arguments with Bohr. I suppose that there is one truism here, in that I can probably believe that my bags no longer exist if they don't arrive at the same airport at which I arrive (and at the same time at which I arrive).

The realistic view is that the wave function, and the system which it represents, have a reality even before any measurement. That is, in quantum mechanics as with classical mechanics, the tree would create a noise (the sound wave) whether anyone is there to hear it or not. Certainly, the differing philosophical interpretations lay at the heart of the many debates between Einstein and Bohr. Which view is correct? Well, there is a significantly large number of physicists who seem to agree with Bohr, which certainly explains why there are so many YouTube videos which talk about the wave function as only a representation of possibilities. On the other hand, it seems that a majority of philosophers lean the other direction, and support a realistic view. In this realistic view, the wave function would have a well-defined representation in terms of a probability distribution well before any measurement is made. Then, the idea that the measurement defines something becomes a view that the measurement process maps the wave function into whatever basis (of possible outcomes) exists for this latter process. But the measurement does not create the realism of the result, only its exact form.

To understand the difference of these two approaches, one can turn to the view described in Einstein's famous paper with Podolsky and Rosen.† Here, in considering their quantum system, they ponder what might happen if one measurement or another, different measurement is made. But, to Bohr, this would constitute two different quantum systems, as each is defined by the measurement that is made. To EPR, reality existed before any

---

*N. Bohr, *Nature* **121**, 580 (1928).
†A. Einstein, B. Podolsky, and N. Rosen, *Phys. Rev.* **47**, 777 (1935).

measurements. To Bohr, on the other hand, reality was defined by what was measured. In essence, this is the extreme limit of Mach's philosophy, as it is no longer possible to think about what might exist before the measurement. Mostly, though, philosophers don't hold with this extreme view, and consider that quantum theory does not rule out the possibility of multiple measurements on a single system. Norton explains this point succinctly by stating[*]: "When we have a spread out quantum wave representing some particle, standard algorithms in the theory tell us what would happen were we to perform this measurement, or, instead of it, had we performed that measurement. Generally the description of what would happen is expressed in terms of the probabilities of various outcomes. But there is no difficulty in recovering the result. Thus, there seems no problem as far as quantum theory is concerned when EPR assert what would happen were this measurement or another incompatible measurement to be performed."

The ideas of EPR reality before the measurement also seems to be supported by the concept of weak measurements that has arisen in quantum theory over the past couple of decades. In this approach, the quantum system is only weakly coupled to the measuring device, just enough to indicate a measurement. But it is assumed that this does not disturb the quantum system to any great extent. This seems to contradict some of the basic tenets of quantum theory, at least as they arise from Bohr's ideas. Yet, it is argued that these weak measurements lie within the boundaries of quantum theory and do not violate any fundamental concepts.[†] This is easily supported if one takes the view of reality that the system has a well-defined wave function, even if it is spread out in its representation of a particle, even before the measurement is made.

---

[*]http://www.pitt.edu/~jdnorton/teaching/HPS_0410/chapters/quantum_theory_completeness/index.html
[†]Y. Aharonov, D. Z. Albert, and L. Vaidman, *Phys. Rev. Lett.* **60**, 1351 (1988).

So, it is clear that the philosophical basis in which one chooses to believe certainly has an influence on just what version of quantum theory becomes believable. But it remains somewhat odd that Bohr's (and Mach's) extreme views have become so ingrained into what has become the common conception of quantum mechanics. We shall return to this discussion later, as it is best to ponder over these points for some time. But perhaps it is a result of the need that Bohr apparently felt to intimidate those who differed with his views.

As mentioned above, we need a level of reality in our everyday life. For example, we develop a microprocessor with a certain set of functional units. When the program is input, it is desired that these units are utilized in a causal deterministic fashion which leads to a desired output. Certainly, there is always the chance of an error, but modern computers have error checking built into the system. We would like to have the same answer appear every time we run the same program. We would not like a probabilistic output that required us to repeat the process a million times to be able to say that this is the likely output. If we are using this microprocessor to compute the flight of our aircraft, in concert with the GPS satellite systems, we certainly don't want the answer from the computer to be "well, that might be the airport we want." Nevertheless, the semiconductor material with which the microprocessor is made is a quantum material. Quantum mechanics governs the properties, especially those that are used to create our transistors and circuits. Yet, we want these quantum systems to give us definitive answers. We are not ready to have the answer to be the "maybe" that Bohr and Mach would have us expect.

So, the philosophical problem is just where reality begins. Is it with the measurement? If so, then I don't have to be in the forest at all, if I leave a microphone and tape recorder. Is there a sound if I never replay the tape? Perhaps, then, reality occurs when I sense the results. Certainly, this seems to be more in line with the statements of Mach. If that is the case, is the resulting reality shaped by the consciousness of the observer? But if reality is with the measurement, then many measurements are made deep within the microprocessor with branching processes defined by the program and based upon those measurements. In turn, this disconnects the measurements from the observers consciousness, as he is only aware of arriving (or not) at the desired airport. Einstein, on the other hand, would want reality to exist even before the measurement, with the latter only relaying just what that reality entailed. Well, we will return to quantum mechanics later, as we have just touched the tip of the iceberg. But you can see that everyday life has certain realities that often conflict with the interpretations of quantum mechanics.

Chapter 5

Some Views of Science

In some sense, the concepts of Mach and Bohr discussed in the previous chapter are separated from the real world interactions in which we live our day to day lives. But then, are those considerations limited to an esoteric corner of science related merely to the philosophical interpretations of quantum mechanics? We will see that this question arises in a much more general discussion of a wide range of sciences, so it is certainly not limited to such an esoteric corner. In this chapter, I will explore these points a little further to illuminate how the question arises elsewhere. While this is certainly of interest, at least to me, some other observations which have appeared to me during the last half-century also will be discussed in this chapter. In some sense, these discussions may more likely be termed rants by me, but they have probably also occurred to others during this period.

Pathological Science

In graduate school, I managed to obtain a copy of a manuscript due to Irving Langmuir entitled "Pathological Science." It was

50 Years in the Semiconductor Underground
David K. Ferry
Copyright © 2015 Pan Stanford Publishing Pte. Ltd.
ISBN 978-981-4613-34-7 (Hardcover), 978-981-4613-35-4 (eBook)
www.panstanford.com

a transcript of a talk he had given in December 1953. This was before the days of the internet and rapid acquisition of PDF files by everyone. So, this copy was a copy of a copy of . . ., and so on. The fact that it was marginally readable just spoke to the efficiency of the Xerox machine in those days (I confess that a copying machine will always be a Xerox to me, just as tissues will always be Kleenex, but I am not alone in this, as the process of a brand name becoming the common usage term is well known). Today, the manuscript has been redone with newer technology and is available on the web as well.* In the talk/ paper, he reviewed several questionable "discoveries" of new scientific effects, some of which he had been invited to witness and discuss with the discoverers. But there were other effects that he described in the talk. The common point was that these effects were not likely to be real physical processes. Rather, they were observations which the experimenter almost wished were real, sometimes in obvious contradictions to accepted scientific principles. These were termed "pathological science," in keeping with the title. He pointed out that such pathological science has a number of common characteristics. In a sense, these effects can be characterized as wishful thinking, as I mentioned. But, as he says, the largest effect that is observed is usually barely detectable and the magnitude of the observed response tends to be independent of the amplitude of the causative agent. In spite of this, the supporters usually claim great accuracy even though the fantastic theories are contrary to the normal understanding of physics (or other appropriate science). Moreover, any criticisms are answered by poorly thought out arguments usually conceived on the spur of the moment.

An important issue here is the role of the causative agent mentioned just above. In classical physics, every observation can be associated with a causative force. But, in the discussion of quantum mechanics in the previous chapter, it was claimed by some that causal behavior is not part of quantum mechanics.

---

*http://www.cs.princeton.edu/~ken/Langmuir/langmuir.htm

Realism lies in the observation of the event, and no reality exists prior to this observation. So, are we then to associate observations of quantum behavior with no connection to the causative agent? This is part of the argument that must be considered in any search for the reality of quantum events.

Most of the effects Langmuir discussed have gone away with time, but some are still with us although often in new forms and claims. One such might be the field of extrasensory perception, which he noted in connection with a program by a noted professor at Duke University. His discussion is interesting particularly as this field of endeavor fits all of his criteria for pathological science. But, as I mentioned above, such programs tend to rise from the dead again and again. In the '70s, while at ONR, there was a program (once again) at Duke and at a major commercial research facility in California to use extrasensory perception as a means of communicating with submarines. The research methods used and the nature of the results were much the same as discussed by Langmuir, so not much new had come into the field since his time. Importantly, there was still the belief that results that gave values below the random expectation could be taken as positive results—it was obvious to the researchers that these participants were fighting against revealing their cognitive abilities and this led to low values lying below the average norm. While this field has lived on, others have come into existence since Langmuir's time. Examples are cold fusion and polywater—new concepts that more or less rapidly appeared and then fell from grace.

Langmuir himself considered flying saucers to be an example of pathological science, but this has not impacted the cult of supporters. Close to home, there was the occurrence of the Phoenix lights, a group of mostly stationary lights that appeared over Phoenix one night near the end of the 20th century.* These were observed by a large number of people throughout Arizona and Nevada, including the governor of Arizona. They seem

---

*http://en.wikipedia.org/wiki/Phoenix_Lights

to have reappeared a number of times since then. The official response is that in each case an air force plane (or planes) had dropped a number of flares at high altitude, so the flares were visible from a considerable distance. Arizona has a reasonably large number of air bases, and a significant part of the southern part of the state composes the Goldwater bombing range. So, the official version has some credibility to it. But, to the believers in UFOs, there remain questions which the official explanation just doesn't answer properly. But, then, Langmuir said this would be the case.

To Langmuir's list, I would add the pursuit of Dr. D's perpetual motion machine vis-à-vis dissipation free computers, as it fits naturally within the description provided by Langmuir. Support lies mainly within a group that resembles more a cult than a reliable assembly of scientists. There are claims of great accuracy, while the physics is contrary to what we have come to understand. Criticisms are met by spur of the moment arguments which always lead directly to their original thesis. Even worse are the concepts put forth by the faithful for new implementations which also have great claims of accuracy contrary to our understanding of physical laws (here I am thinking particularly of the billiard ball computers). Within the definitions described by Langmuir, one can readily feel comfortable including this within the group of strange beliefs.

Having brought back a discussion from the last chapter, it seems to me that we also could come back to the Mach/Bohr philosophy via some studies from biology. We recall that Mach and Bohr believe that the existence of processes, such as quantum systems, depends upon the presence of measurements which lead to our conscious awareness of the results of those measurements. Langmuir included in his paper a discussion of *mitogenic rays*, which addresses the issue of whether or not plants can communicate. Supposedly, mitogenic rays are emissions given off by living plants, and several hundred publications appeared describing the effect in the 1920s and 1930s. It was claimed that these rays would travel through quartz, but not through glass, which would indicate that they may lie in the ultraviolet portion of the spectrum. Indeed, it was generally believed that the radiation lay in the 190–250 nm wavelength spectrum with an intensity of a few to a few hundred photons per second per square centimeter. A careful study done by the National Research Council in 1937 found no evidence for any such rays that could be detected by either physical or biological sensors of the day. Nevertheless, the Soviets continued to work on these rays long after Langmuir's discussion. But this was before the new age movement and the onset of the great consciousness, which brings us to the heart of the Mach/Bohr ideas. Today, we are told that you can indeed communicate with your plants if you will only treat them as equals. What is the basis of this communication? Could it be that we, as well as plants, emit and detect ultraviolet radiation? We are told that even those of us who do not claim great psychic abilities should be able to approach a large tree and pick up signals of the mood and personality of that tree. Science fiction goes somewhat further, where we have the Ents from Tolkien's *Lord of the Rings*. But there has always been a claim for sensing something from the plants in ancient spiritual religions where the high priest or shaman would be capable of such processes.

But then perhaps the early workers and the National Research Council were looking in the wrong part of the spectrum. More recently, it has been reported in the scientific literature that plants might be communicating with one another through sound waves. The evidence seems to be that plants grown close to one another grow better than when they are grown in isolation. I wonder what Langmuir would say. Then again, I wonder what Mach would say. If, in Mach's world, sound doesn't exist until it is experienced by our consciousness, how can it exist to allow communications between plants? Can plants have the consciousness necessary for such communications (which presumes that one plant must measure/detect the emissions from another plant)? Perhaps we are being too critical in applying the criteria for just who (or, perhaps it is better to say what) experiences the measurement of the sound. But then other scientists have claimed to discover that plants communicate through an interconnected root system, much like our own internet. All of this boggles the mind. Are we to infer that there is a state of consciousness within these plants that allows them to recognize such signals, whether by mitogenic rays, or by sound, or by some process in the roots? If we accept this, it would seem to clearly interfere with the belief that consciousness is limited to humans (or perhaps we should say animals in general). Or, perhaps we should not put too much faith in what we read on the net. If consciousness and faith have entered this discussion, perhaps we should query the leaders of the world's religions as to what all this means. Or, maybe we should just submit that the ideas of Mach and Bohr may be just another example of pathological science.

Following the Leader

Scientists in general, and I guess physicists in particular, seem to have a herd mentality, but I suppose that they are not all that different from the population at large. There is always a top few percent who populate the generally creative and innovative group of scientists. But the rest seem to move from one topic to another as a herd seeking the next feeding opportunity. I suppose this is just human nature, but there are some herd activities that can be downright dangerous, much as lemmings following their leader off the cliff. In the study of small structures such as nanostructures, there has been a belief that everything could be explained by the so-called mean-level separation which was determined by the physical size of the structure. Or, there was a common belief that opening a quantum system to its environment would completely wash out any quantum effects. I have already talked about the error of this latter view in a previous chapter, but there are numerous experiments which have shown the fallacy of both of these views. Nevertheless, far too many people still cite these views without questioning them, and this continues to hinder achieving a proper understanding. The best that one can say about such beliefs is that they have set back the understanding of physics by only a few decades. Some, as we will get to later, may have set back the understanding by at least a century.

It is usually easy to spot the movement of the herd by the appearance of new discoveries that illustrate new understanding or possibilities. The most obvious place for me is the March meeting of the American Physical Society, which is the annual meeting for the condensed matter physics crowd. There is always a couple of thousand members of the herd trying to crowd their way into the topical sessions on the latest greatest breakthrough. Most could have saved their travel money just by reading the research papers of the speakers, which certainly have been published long before the meeting. But I guess there is a certain internal need among them to say that they were there; they heard the great talks before the awards began arriving, or perhaps before it is discovered that the awards are not deserved. The Society cannot be held responsible for either the contents or the herd, as the session topics are proposed by active physicists, who should be aware of the value of the talks that will be given.

There is a website out there that will give you your ranking as a physics author as a function of time. For example, if you see a ranking for a given year as 40%, then it means that you are in the top 40% of physics authors. I have to admit that there is not much information available on just what goes into these rankings, and they may not be as reliable as those for universities mentioned in the first chapter. In any case, when I leaped into the upper half while still in grad school and still today rank in the upper few percent, I have to ask just what all those other members of the herd are doing with their time. It's obvious that they aren't publishing their research (or, maybe they are, and this is more is a reflection on the overall quality than about me). My students would tell you that at this point I would say that to consider this properly, it is time to go and have a margarita and *then* ponder the meaning of such ratings.

Following the leader can have many other philosophical factors, especially for those who "think outside the box," and go a different direction. This can have many aspects. For example, I have known a few physicists who always conceived of the truly difficult problem upon which to work. Much money was spent building nice facilities, but nothing ever came from the projects. Their colleagues always explained this with "But he attacked a really difficult problem." It seems to me that you can do this once or twice, but it is really hard to make a living following such a thread. To the contrary, Einstein is known to have spent his later years working on the unified field theory, trying to bring all the four major forces together in one theory. He is said to have justified this on the grounds that his fame was already made, and a young scientist couldn't afford to spend his time working on this problem. To explain a little further, here is a famous scientist working hard on a major unsolved problem, rather than following the herd. Perhaps there is something to be learned from this.

Sometimes the herd mentality and the lemming theories can lead to a problem of just evaluating the data. Far too often, the researcher just knows what the data is supposed to say, perhaps because he has read some interesting papers on the subject. There are too many examples in fields ranging from sociology to medicine and even to physics, where the data has been cooked (o.k., only selected data may have been used, but in a number of cases it was manipulated) in order to support a predetermined theory. In at least one case, even after the data was exposed to be wrong, it was still said, "But his conclusions were correct." On what basis can one really believe this? In contrast, we should perhaps reward those who fight against such temptations. I have had a Nobel laureate tell me about missing an important discovery because he was not seeing what the data was telling him, so he didn't believe in publishing it. It didn't become clear until he saw someone else's breakthrough paper (which also led to a Nobel Prize). As he put it, he had concocted a number of theories, but he knew that they just didn't fit with the data properly. To me, his genius was in not publishing the theories, since they didn't work. Far too often today, one is led to publish a calculation or a derivation just because it has been done, not because it leads to a new or revised understanding.

In my youth I went to a conference on superconductivity that was held at Stanford University. This was almost a decade after the famous theoretical paper of Bardeen, Cooper, and Schrieffer defining the basis for superconductivity, but before they received the Nobel Prize. This paper made a number of important predictions and contributions to the theory. Yet, the keynote speaker, a famous experimentalist in superconductivity, made the astonishing statement that all the theorists in the world had not raised the transition temperature by 1 mK. Now, the transition temperature is the point at which a normal metal becomes superconducting (zero resistance) as the temperature is lowered. At the time, the highest known value of this quantity was just below 30 K (around -400 Fahrenheit). In a sense, the speaker was correct, because all of the materials science which led to improvements in this quantity was done by experimentalists. Yet, there was a herd of theorists working frantically to try achieve a higher temperature, yet being just as unsuccessful. It was not for another couple of decades before Alex Müller and Georg Bednarz, working at IBM Zurich, achieved a breakthrough in their experiments on barium doping of some metallic oxides. This opened the field of what is called *high temperature superconductivity*. Sure enough, the herd arrived and the hype machine went into overdrive with predictions of a great many applications which would revolutionize the world, including dissipation free distribution of power. At about the same time, there were predictions that such metallic oxides might be superconducting via a new interaction, which was ultimately verified. Unfortunately, the experiments were done independent of this theory (at least in published form), so technically the conference statement had not been overturned. And, in spite of the herd working the problem, most of the hype and claims have not borne fruit, and the transition temperature has only been raised to a little over 100 K (so, that it remains near −220 F). But this is good enough to now be able to use liquid nitrogen cooling instead of the more expensive liquid helium. Whether further advances will be forthcoming is a question for the future, but the old situation has returned with a herd of theorists working actively and arguing over details, but not affecting the transition temperature.

Sometimes the herd doesn't arrive until after the hype has been launched. That is, too many are reading the press releases instead of the scientific papers. One such area is topological insulators, which have been all the rage the past few years. These materials have exotic and very interesting properties. One key property is that these materials often possess a Dirac-like band structure (bands which cross linearly at a point and which have no energy gap; graphene is the prototypical fruit fly for such a material), which is topologically stabilized at a surface or interface, while the bulk material is normally an insulator. In a normal semiconductor, one can change the energy gap between the conduction and valence bands by the application of pressure and by very large electric fields. That is, the properties can be modified by the application of a number of external forces. In a topological insulator, this cannot be done—the topology that produces the Dirac-like bands stabilizes the bands against such forces. One prototypical material is a heterojunction between the negative-bandgap material HgTe and the positive-bandgap HgCdTe or CdTe. In HgTe, the energy state that composes the normal top of the valence band and the one that normally is the bottom of the conduction band are inverted. When, crossing the interface between the two materials, this band ordering must reverse itself, and this can lead to the Dirac-like band structure at the interface. The question here is why the importance of these topologically stabilized bands was not appreciated by theorists (or experimentalists) much earlier. The nature of bands in the HgTe/CdTe superlattice were being discussed at conferences at least as early as 1979. So, the nature of the energy structure in the interface between the two materials must have been known. In retrospect, there may well have been two reasons that the interesting interfacial properties were not appreciated. First, the concepts of topology were not in the mainstream of condensed matter and materials physics at the time. Secondly, the intriguing ideas of Dirac-like bands in materials was not appreciated at the time, as graphene has only been isolated in the last decade, even though its interesting band structure has been known for many decades.

In one sense, topology crept into the field with Laughlin's introduction of a structure and phase rigidity to explain the accuracy of the quantum Hall effect in 1982. It became stronger with the discussion of fractionally charged quasi-particles in the fractional quantum Hall effect. In the latter, the fraction of available electron states that are filled in a magnetic field is the ratio of the number of electrons to the number of flux quanta in the structure (the flux quantum is defined as the ratio of Planck's constant h to the electric charge e). Thus, in the $v = 1/3$ state, there are three flux quantum for each electron. When the magnetic field is exactly at this filling factor, all the states in this level are completely full. If we now add a single flux quantum, then the unoccupied state must hold a quasi-hole with a charge of 1/3 to fit the statistics, a fact that has been verified experimentally. The interaction between two fractional states has shown the phase interference that theorists believe could be adapted to created some fascinating new types of logic gates for computing. So, topology and topological computing came to the forefront with these fractional charge states, and the genie was out of the bottle. As mentioned, appreciation of the Dirac-like band structure came later. But it was only with the isolation of single layers of graphene that experiments could begin to establish the unique properties of this band structure. And, studies of the quantum Hall effect quickly illustrated that these bands had different properties than normal semiconductors.

So, by 2005 or so, there was a reasonable appreciation of the role of topology in condensed matter physics, and there was a demonstrated difference in behavior in Dirac-like bands. At this point, the stage was set for someone to suggest a more general view of topological insulators (although the idea of a topological superconductor is older), and the ideas on topological changes were widespread and growing a few years earlier. In spite of this nice argument, the phrase "topological insulator" actually seems to have originated in biology, but burst into condensed matter theory at about the same time. Today, the focus is more directed at materials (and structures) that have insulating bulk properties, but exhibit metallic surfaces. It is only now, with the discovery of a few additional materials with this property, that the hype machine has switched into high gear, and we are awash with predictions of a great many applications which are sure to revolutionize the world. As expected, this led to the arrival of the herd, but so far not to any great breakthroughs or earth-shattering new properties.

What Goes Around Comes Around

A number of agencies in Washington like to rotate their program managers every few years. This is thought to be a method of bringing new "blood" into the agency, as the incoming managers are supposed to be full of new and modern ideas. This may actually be the case on occasion, but it is also a way of making sure that the corporate memory is erased every few years. Earlier, I mentioned extrasensory perception, which Langmuir labeled as pathological science, but which reappeared a couple of decades later as a method of communicating with submarines. When the corporate memory is reset, ideas often come around again. Each time they reappear, it seems that they are hailed as the latest, greatest idea. They are predicted to give fantastic new approaches to whatever field is being discussed at the time.

Recently, the idea of probabilistic computing has become a novel new concept. Of course, this promises faster computing with lower power, although one may have to accept "maybe" as an answer. Moreover, there is the belief that this approach, via analog processing, might prove to be the wave of the future. As you might guess from the title of this section, none of this is new and/or surprising. It's all been here before, and there are several different scenarios that have appeared over the years. Some of the names that have been applied to this idea are fuzzy logic, stochastic computing, neural networks, biological networks, and so on. It is important to say that some of these have found useful homes, though for other reasons than just probabilistic computing. All of these are basically more modern versions of the analog computer, as they mostly use analog processing somewhere within their architecture.

The analog computer has an old and rich history, apparently dating back to the ancient Greeks. Electronic versions were created prior to world war I for fire control computers, and have evolved from there. The ones I used in my student days had the active elements created from operational amplifiers, which could be programmed as integrators. Thus, these analog computers could rapidly solve integral equations, so long as they were not too nonlinear. In the '60s, Lotfi Zadeh, a professor at Berkeley, began to develop his ideas of "fuzzy logic." Here, instead of the binary logic developed for digital computers, he would use analog gates to introduce a probabilistic basis for the computation. This even was applied to quantum mechanics, already a probabilistic approach, with the use of non-orthonormal basis sets. Zadeh's approach has sometimes been called possibilistic computing, and it gained considerable traction during the '70s. But the defense department gradually lost interest, since the applications generally have to have a definitive "yes" or "no" answer, rather than a "maybe." If one needs to ask why this is the case, then he should google "USS Vincennes" and its tragic error in 1988.

In the '70s, probabilistic computing had grown into a sizable program at Illinois under Prof. Poppelbaum.* This was now called stochastic computing. The basic idea, particularly with the connection to neural nets and cellular automata, dates to a talk given by John von Neumann in the late '40s. The claim is that stochastic computing is inherently robust against noise, and is faster than normal binary computing, but at the expense of accuracy. The weakness lies in its inherent randomness. There is still evidence that this approach is particularly useful in decoding error correcting codes, so there is some applicability even today. In a strict sense, stochastic computing is not analog in nature, but evaluates mathematical operations through probability rather than with arithmetic.

Neural networks and cellular automata are quite similar in concept. An individual logic "gate" has a set of inputs, and from this set, an output level is determined. In neural nets, the decision is based upon an analog function, and both the inputs and outputs are analog signal levels. In cellular automata, both inputs and outputs are digital levels, and a decision function is used to map from the inputs to an output. Hence, it is the neural network which becomes our analog computer. As mentioned above, the idea for this comes from von Neumann, who was inspired by the McCullough-Pitts computational/electrical model of the neuron. During the '50s and '60s, most of the development was confined to the computer science and artificial intelligence community due to the recognition that a learning algorithm could be used to refine the strength of the interconnects among the artificial neurons. The use of these networks in electronics burst upon the scene in 1982 when John Hopfield published a paper on the applications for associative memories. Artificial neural networks then became the "fad" topic with its large herd following. We

---

*P. Mars and W. J. Poppelbaum, *Stochastic and Deterministic Averaging Processors* (Peregreinos, New York, 1981).

were among those drawn to the field, since it was clear that semiconductor integrated circuits were a fruitful approach to mass produce useful networks. One of the difficulties is that each of the inputs to the processing cell must be multiplied by a weight that is assigned to that input. We recognized that this was a natural function for a MOSFET, and developed several circuit layouts for learning in layered neural networks (layering reduces the need for total interconnection among the neurons; instead they are organized in hierarchical layers), many of which were actually implemented in silicon.*

One of the important limitations to artificial neural networks is in the operation. Normally, integrated circuits are clocked, so that operations occur in a synchronous manner. In concept, there is no reason against operating the artificial neural network in an asynchronous manner without a clock. In general, this is thought to be more in line with how the body operates, as nerves give signals and the brain processes these without an obvious clock. One of our studies was the role of this asynchronous behavior on a neural net with threshold decisions in the neural core. That is, a particular set of input states was mapped into a variety of different output states according to a defined probability distribution. Here, it was found that this led to the creation of random outputs in the system, with a number of instabilities.[†] This was not exactly what one would want in a central nervous system nor even in a well functioning integrated circuit. Most circuit designers take umbrage at having random behavior within the chip.

Even with synchronous clocking, some disconcerting events could be found when treating the analog neural net as a memory. As voltages are switched on the interconnection lines running through the array, these could influence the state of a particular

---

[*]L. A. Akers, M. Walker, R. Grondin, and D. K. Ferry, "A Synthetic Neural Integrated Circuit," in *Proc. IEEE 1989 Custon Integrated Circuits Conference*, paper 12.6.1.

[†]R. O. Grondin, W. Porod, C. M. Loeffler, and D. K. Ferry, *Biol. Cybern.* **49**, 1 (1983).

cell through stray capacitive coupling. With a digital cell, this is not important as the cell is forced into either the 1 or the 0 state. With an analog cell, however, these couplings lead to random variations in the analog level of the cell. This can lead to upset and random signals coming from the cell. In actual fact, designers worry about nearest neighbor interactions in memories through a general process known as device–device interactions. We defined a wide range of physical properties which were subject to these effects, either classically or quantum mechanically, in real integrated circuits as the size of individual cells became smaller through scaling. In fact, the superlattice effects discussed earlier fall into this category of physical effects.

There remained significant interest in the artificial neural nets for quite some time. This was, as mentioned, due to the ability to use learning algorithms to adjust the different weights to have the networks optimize themselves to perform the desired task. Nevertheless, over time, digital computers became ever more powerful, and digital algorithms became better and faster for use in artificial intelligence applications. Thus, the analog neural networks gradually faded out of the mainstream. But some probabilistic computing remains, particularly in the world of quantum computing (to which we return in a later chapter). But, as mentioned, a "new" paradigm of probabilistic computing with analog elements seems to have re-emerged this century. Perhaps this has appeared with the loss of corporate memory in Washington, or perhaps it is really a new concept. Time will tell.

The Sky Is Falling

I recently noted an insightful article whose major point is that the United States is losing the lead in nanotechnology. In keeping with the spirit of the previous section, it seems to me that about every five years, we see the appearance of several such articles talking about losing the lead in ". . ." where you can fill in the blanks with your favorite technology. Of course, one must ask the question as to whether or not we ever really had the lead. In many fields, particularly nanoelectronics and integrated circuits, we certainly do have the lead. In fact, it is questionable if anyone else was ever able to even stay close. But what constitutes a lead? How are we to evaluate who is leading?

The fact of the matter is that the flood of people who want to come to the United States is not abating. Whether they come seeking jobs, education, opportunity, or whatever, the demand is growing. If we judge by the number of foreign students, then university education in the United States is second to none (regardless of anybody's surveys indicating otherwise). These students are voting with their feet, and that is the most important measure as they are pinning their future hopes upon this education. It is important to note that graduate education is largely based upon the capabilities inherent in frontline research programs. It often is these research programs that lead to the reputation of an institution, and it is this reputation that draws the students to the programs. The state of California, by itself, has more major research universities than most foreign countries. One might think that if we were losing the lead, some of these students might redirect themselves to wherever it was that took over the lead. The fact that no other country, or region, has slowed the exodus to the United States, especially in nanotechnology, might suggest that we have not lost the lead, at least in this measure.

To create a massive industrial base for any given technology not only requires large industries in place as the technology arises, but also a climate that provides for new industries to arise and grow. It seems to me that no country other than the United States has the entrepreneurial environment based upon innovative development of new ideas, and the capital investment firms that are just as important. Is there a research university in the US that is not surrounded by multiple incubators designed to ease the startup woes of new companies? At the same time, how many foreign research universities see these same incubators? Of course, this is the present state, and things can change rapidly, especially when national governments get involved.

Graphene was discussed above as an exciting new research area. We recall that graphene is a single layer of carbon atoms which can be peeled from graphite. It was only isolated a decade ago, and has been the center of attention for major research efforts around the globe. It is true that only in Europe have large national investments been put into research facilities to pursue the results of this discovery. For example, the British government has poured a significant amount of money into a large facility at Manchester University, where graphene was first isolated. This has, of course, attracted a great deal of attention worldwide, and perhaps might indicate that we have lost this technology to the British. But I wonder if this is really the case. In early 2014, the Gates Foundation awarded a sizable contract to the Manchester research facility, which may be an indicator. However, the contract was to develop graphene-based condoms, hardly advancing a frontier technology. Is the decision of the Gates Foundation to fund this program outside the United States a sign of the decline of our nanotechnology, or is it a sign that only a foreign institution found this an exciting area of scientific research? Now, I have to admit that such a project invokes lots of questions about how to conduct the testing program. Will there be miniature versions for lab animals? Or, will they recruit student assistants to test the various lab samples that are developed? Perhaps they will use their national health service to recruit participants for a large survey to test their new products, which is an approach often used in medical research. I can just imagine the advertising.

Chapter 6

Science and Life May Be Fickle

Most of us are familiar with how the same object may be described quite differently in different countries. Even when the countries speak nominally the same language, items will be described quite differently. For example, in my car, the spare tire is kept in the trunk. But to others, a trunk is a very large piece of luggage. An Englishman keeps his extra tire in the boot. To me, a boot is a piece of western footwear. We could go on this way forever, but this gives us enough to make the point that context and source of a remark or question greatly affects the meaning.

In keeping with this idea, it was remarked at various points that the manner in which a question is asked can be quite important. The connotation of the question is almost as important as the question itself. In starting this chapter, I want to come back to this point and describe how this can lead to very different outcomes. Then, I will go on to discuss how I came to be a theorist on occasion and how theorists can do great damage at times, in line with the law of unintended consequences.

50 Years in the Semiconductor Underground
David K. Ferry
Copyright © 2015 Pan Stanford Publishing Pte. Ltd.
ISBN 978-981-4613-34-7 (Hardcover), 978-981-4613-35-4 (eBook)
www.panstanford.com

Asking the Question in Context

Asking a question in which you expect to get a single very informative answer can be quite difficult. Earlier, I talked about a visit with FA18 pilots at Miramar, in San Diego. These were marine aviators, but probably typical of any combat pilot. They are extensively trained, and highly skilled, in the ability to get the job done with the available equipment. So, when one of our party asked how we could improve the equipment to better enable them to get the job done, the answer was obvious—they could get it done with what they had. Of course, this was useless to us, as our task was to try to recommend new electronic capabilities to DARPA. In such a scenario, it was clear that pulling the information from them would be difficult. I suppose that, given enough time and adequate amounts of liquid to break down the natural reluctance, we could have gotten to suggestions on improvements. But we didn't have the time or the lubricants. Under the right conditions, it is clear that one could have gained a lot of information on what improvements would be useful for the next generation of aircraft. The point is just that, in some circumstances, asking an ill-conceived question is no more than small talk. We can illustrate this in another way.

Richard Benson, another educator, has written some books on the funny answers he has received on tests. One incident in particular makes the case. The question asked was "What is a vacuum?"* The student's answer was "Something my mom says I should use more often." Now, in many circumstances, such as a class in home economics, this answer may be totally acceptable. It certainly is not a wrong answer—I heard my wife raise the same issue when my daughters were teenagers (truth be told, I can remember my wife raising the same issue with me). But, in a class in chemistry, it is totally out of context as an acceptable answer. On the other hand, if the question is taken out of its

---

*Richard Benson, *F in Exams: The Very Best Totally Wrong Answers* (Chronicle Books, San Francisco, 2011).

context (of chemistry), then there is no way to jump to the conclusion that this answer is unacceptable.

There is another story that explains this idea of context even a little better. The story arises about a physics test in which the question was: "You are given a fine barometer; how would you use this to measure the height of the XYZ building."* In the story, the student gave an answer different than that desired by the instructor, and was graded down. But the student protested that there were several answers possible for such an ill-posed question. Rather than repeat the story, I will paraphrase it in terms of context, all of which relates to the material for which the instructor was testing. If the course was an introductory course on length, time, and gravity, than one acceptable answer might be to tie a string to the barometer and lower over the edge from the top of the building. By measuring the length of string, one would immediately know the height of the building (which was the answer the student submitted). On the other hand, one could throw the barometer off the top of the building and measure how long it took for the barometer to hit the ground. By some simple equations from physics, one could then determine the height of the building with the well known expressions for gravity (and even correct for air resistance).

If this were a course in trigonometry, an entirely different approach is warranted. Here, one could place the barometer on the ground, and then measure the length of the building's shadow and the length of the barometer's shadow. Using the properties of equivalent triangles, one could then determine the buildings height from the height of the barometer.

If the course were in the field of negotiation and business, one could give an entirely different answer. The student could take this barometer to the building superintendent and open negotiations with the ploy: "Look at this magnificent barometer.

---

*The story appears to originate in an article by Alexander Calandra in *The Saturday Review* in September 1968, but has been elevated to an urban legend in physics.

I will loan you this wonderful device if you will tell me the height of this building." Clearly, all of these answers are correct when the context of the test is clearly delineated.

But one has to admit that the course topic over which the test was administered was likely a study of atmospheric pressure and its variation with altitude. In this context, simple measurements of the barometric pressure at street level and at the top of the building would yield the expected answer to the original question. After all, this is one purpose for the barometer and it serves as the basic altimeter in aircraft, at least it was, when I was flying years ago and I think that it still is albeit modernized over the version familiar to me. At that time, it was a good guess at the altitude, because it always had to be adjusted to the known altitude, and barometric pressure, of the airport prior to takeoff. There was no guarantee that it would be calibrated properly at any other location. In the original story, the student had even more solutions than what I have given here.

The Gospel of the Laboratory

When I was working on my doctoral degree, I actually did the work in a laboratory that was primarily devoted to plasma physics. In general, the experiments in this laboratory were large, room-sized endeavors with high voltage, and high power, required for parts of the operation. In contrast to this, my investigation of semiconductor "plasmas" was small and much easier to operate. But the interactions with the group in this laboratory provided a number of lessons and reasons for remaining in the field of semiconductors. One reinforcement for my choice to work in semiconductors came just after I had joined the laboratory.

The experiment was a plasma "pinch" experiment. In this experiment, a large bank of high-voltage capacitors was used to store a reasonable amount of energy. This would be discharged over a very short period of time into a single turn coil of copper that was wrapped around the glass tube in which the plasma was situated. The induced currents and magnetic fields in the plasma would cause it to compress into a single filament down the axis of the tube and coil. This compressed plasma was then examined by a variety of microwave and laser probes. The experiment was currently the domain of two graduate students, who I shall call J and R. During each pulse of the plasma, the entire capacitor bank was discharged into the coil. This was achieved by using a large number of low impedance coaxial cables, one from each capacitor, which attached to the coil by a copper header. On this particular day, J remarked that he heard what sounded like a small spark somewhere near the header. So, R suggested that J get down close to the coil. They would turn out the room lights and then discharge the capacitors, and perhaps J could see just where the spark originated. So, they proceeded in this vein. Now, the header is composed of two copper plates (which are in turn the leads to the coil) which are separated by a plastic insulator. If the insulator fails, then the capacitors discharge through a short circuit which would have catastrophic results for the main part of the insulator. Indeed, as they put their plan of action into operation, the insulator exploded in front of J's eyes. Fortuitously, he was protected by the coil itself, and suffered no more than ringing in his ears from the explosion. Most of the parts of the insulator went out the ends of the coil and away from J. The energy imparted to the parts of the insulators was sufficient that one piece hit the wall of the room, rebounded through the open doorway and was then found a few hundred feet down the hallway. As the dust settled in the room, R leaned around the equipment cabinet to face J and asked "Did you see the spark?"

Needless to say, observing such activities reinforced my commitment to working with semiconductors. You might say it even embedded in my mind at this early stage that I should work with nanostructures, where the danger to life and limb was much less. As I remarked, however, there were other lessons to be learned here. One of these led to what I call the gospel of the laboratory.

A second large experiment was concerned with what are called ion cyclotron waves. In the ionized plasma, the electrons are separated from the nuclei, so that each can feel the presence of an external magnetic field. The experiment involved a long glass tube in which a low density plasma was created. This tube was encircled by a large number of large coils in order to produce a magnetic field directed along the axis of the tube. Then, once the plasma was formed in the magnetic field, microwaves were propagated down the tube in a manner that the circular polarization of the microwaves would interact with the ionic circulation of the nuclei ions around the magnetic field. Well, on this particular day, the graduate student had been working on the system and a variety of tools lay about. Unfortunately, the professor came into the laboratory and, without checking, turned on the magnetic field to see if it were operating properly. Tools are often made of metal, which is often magnetic, as was the case here. A relatively large wrench was launched down the narrow space between the glass tube and the magnet coils, rattled against each as it moved. One could literally see the fear grow in the eyes of the student. If the glass tube were broken, it would mean months of down time before a new tube would be in place. Fortuitously, the tube survived until the power was turned off. In discussions with the student after the fact, I gained an awareness of just where he thought the proper place was for the professor, and it didn't involve being in the laboratory. One byproduct of this was that, when the professor visited my part of the laboratory, he kept his hands in his pockets. My own work also involved glassware (dewars and so on) and large microwave instruments.

Many years later, as I was setting up my own laboratory as a young new faculty member, I had already taken on a couple of students. Mainly it was the students who were doing the manual labor of putting the experiments together and doing various tests. One day, I came into the laboratory when the senior graduate student was there. As I reached for one of the switches,

I noticed the same fear appear in the eyes of my graduate student as I had seen in my colleague who was working on the ion cyclotron resonance experiment. Good grief! Was I becoming my professor? Was I now seen as a danger in the laboratory, and as one whose presence was an inhibitor to the progress the graduate student hoped to make? Now, I was one who enjoyed doing experiments. But it appeared that I had transitioned almost overnight into a hazard to be prevented from coming into the laboratory. Yet, in thinking about this transition, it became clear. I could not keep up with the daily flow of problems and solutions that appeared in the laboratory due to the other time eating tasks that faced a professor. There were classes to create and teach, undergraduates to mentor and assist in design of their own project laboratories (that I mentioned earlier). There was plenty of time to interact with the graduate students and discuss the experiments and analyze the data. But it was clear that I could no longer be involved in the daily operations. From these considerations emerged the gospel of the laboratory.

It would become important to designate a senior graduate student, who essentially was the one closest to graduation. He would become the leader of the laboratory. With that designation also came the responsibility to train new students as they joined the group. This part was important, as he had a very good reason for doing this training well—he didn't want the new hands to screw up his experiment. Thus, new people were trained in running things in the lab as he wanted them done, not as I may have wanted them done. So, I had to make sure that his view of operations more or less coalesced with my views.

Years later, I accompanied a pair of my students to Japan to use the low temperature facilities at RIKEN and to work with Jon Bird. As we were all sitting in the lab one evening, they asked why I never came into the lab to do experiments with them. First, I explained the catalog description that getting a doctorate was mainly proving the ability to do *independent* research. But then I explained the gospel to them. This they could understand, but it became even more clear to them when Jon and I screwed up late another night and blew all the liquid helium out into the atmosphere. They never invited me into the laboratory after these adventures. And, when Jon accepted a position at Arizona State a year or so later, I explained the gospel to him. This kind of shocked him, as he was not ready for his students to lock him out of the laboratory (as essentially happened some months later).

I used to think that I was unique (don't we all) with the gospel of the laboratory. But one day while the Defense Science Research Council was meeting in La Jolla, I was talking with a colleague from CalTech. He remarked that he had a student finishing his doctorate and had to go back to Pasadena that evening for the "passing of the bat" ceremony. Intrigued, I inquired about this, as I knew he had been an excellent experimentalist as a graduate student and postdoc. He said that his senior graduate student was designated as the keeper of the bat, and when he graduated there was a ceremony to pass the bat to the next student in line. At this point, I was wondering just what the bat was. As he explained it, the bat was a real baseball bat, and the purpose of this bat was to keep him from messing with the experiments. His students went so far as to make a small box with a button, a knob, and a light on the surface. When he pushed the button, the light would come on, and the brightness of the light was adjusted with the knob. Now, this box was kept near the entrance to the lab, and these controls were the only things in the lab that he was allowed to touch. So, apparently, the gospel of the laboratory was far more widespread than I had imagined.

While the developments that followed under this gospel were advantageous to all, they did frustrate my desire to do some of my own research. Since the students didn't want me in the lab, I spent a fair amount of time in my office and this led to an appreciation of theoretically based research, especially in an area called computational science. This developed rapidly, and I soon had several students working in this area. Sure enough, as soon as they understood what they were doing, I was more or less smoothly removed from their computational efforts—just another version of the gospel. I suppose it was just as well, as I could now pursue my own interests while, at the same time, keeping up with both the theoretical and experimental efforts of the students through regular meetings with them. But once I took up the theoretical work, there was a change in the students in the laboratory. It appeared that they achieved a realization in their minds that they were right in keeping me away from the experiments—theorists are dangerous around experiments.

The Danger of Theorists

Apparently, the famous Wolfgang Pauli, a noted theoretical physicist had the reputation of being able to break experimental equipment merely by being in the vicinity. This came to be known as the *Pauli effect*. Now, I never had the opportunity to meet this famous man, but I have experienced the same effect caused by one of my colleagues who also was a pure theorist. Prof. J and I collaborated for many years, so I had a number of chances to observe this strange behavior. Plus, like Pauli, he seemed to revel in this effect and so would tell me about other occurrences that I did not observe first hand. I suppose he knew that word would spread and I would hear about it eventually.

The first time I encountered his strange effect was at a NATO Advanced Study Institute (ASI) we organized in Italy. Over the years, we organized several of these ASIs, but quit after the fall of the Soviet Union. NATO seemed to want to bring in organizers from eastern Europe, which was fine, but then they decided that any new Institutes should be held in the east. Our view was that there was no Italian food in eastern Europe, so there was no reason to organize an ASI there (all of ours were held in Italy, for obvious reasons). Anyway, our first ASI was held in Urbino, Italy, at a small conference facility outside the town. When the day came for Prof. J to give the first of his lectures, he strolled to the front of the room and switched on the overhead projector (this was well before the advent of computer presentations). The bulb of the projector immediately destroyed itself. But we had prepared for this eventuality by providing a second projector. So, Prof. J moved to the second projector and switched it on. Once more, the bulb destroyed itself. Now, this is somewhat beyond the normal expectations of probability, so he felt that perhaps the plug had come loose. He got down on his knees so that he could reach the plug and gave it a thorough wiggling to make sure it was well seated in the socket. At this point, the air conditioning and all the air handling equipment in the facility died. A few moments later, the lights in the room also died. The staff found us new bulbs for the projector and these were inserted just as the power returned, but Prof. J was no longer allowed to use the power switches for the equipment.

I suppose we should have expected this to happen. On our way to Urbino, Prof. J and I had gone to Bologna to see Carlo Jacoboni, a professor in Modena and one of the organizers of our ASI. After a nice lunch, we made ready to move on to Urbino. Just before leaving, Prof. J felt the need to use the bathroom. And then we were off. After arriving in Urbino, Carlo had an emergency phone call and had to rush back to Bologna. Apparently, the toilet had failed to stop the running water and there was a flood in the apartment. This should have alerted us to the likelihood of future problems.

Following the Urbino episode, there was a move afoot to provide Prof. J some punishment for his affliction. He was to give the final lecture at the ASI, and this would follow a coffee break on the last afternoon. During the coffee break, he retired to a small room at the rear of the auditorium to go through his viewgraphs. I followed him back there and a discussion followed. My thrust was that it was highly unlikely that anyone was going to return after the coffee break. They would just head to the bar as it was already late in the afternoon. His reply was surely they would come back for his excellent lecture. So, as the time for him to speak approached, he gathered up his viewgraphs and moved to the podium in the auditorium. Sure enough, the start time came and went with no one entering the auditorium. After five minutes or so, with still no audience, he wrote on the board "wife, three children and 10,000 small devices to support; give freely," then sat down on the floor with his back to the board and lit a cigarette (also well before smoking became unacceptable in polite society). We talked a little longer, and finally after some fifteen minutes, the audience started filing in and taking their seats. Then, someone noticed the comment on the board and 100 Lira coins* began to fly out of the audience toward the good Prof. J. I think that, in the end, he turned a nice profit, but clearly he was shaken and his lecture suffered accordingly.

A few years later, I heard about the visit of Prof. J to Bell Laboratories to visit a colleague. Now, this colleague was a leader in the area of heterostructures grown by molecular beam epitaxy, so he had a relatively large laboratory occupied by these large machines and the necessary characterization equipment. Our colleague was (foolishly) showing the laboratory to Prof. J, when he was called away for a moment. But he warned our friend to not touch anything. Nevertheless, while Prof. J claimed to just be reading items on the bulletin board in the lab, alarms sounded, lights started blinking and all of the equipment appeared to be going into emergency shutdown. Our colleague came running back into the laboratory shouting "What did you do?" The reply that he was just reading the bulletin board did not carry much credulity, and he was ushered out of the lab in a quite unceremonious manner. I do not believe that he was ever allowed back into any part of Bell Laboratories.

I would like to believe that this is not an affliction shared by all theorists, but I am not so sure. It certainly has resulted from my being near the equipment, but in not such a direct fashion.

---

*At that time, 2000 Lira was about the equivalent of a modern 1 Euro coin, so 100 Lira was a relatively small amount.

Earlier, I remarked about taking a pair of students to RIKEN to use the laboratory when Jon Bird was still there. After a day of work in the laboratory, the four of us went out to a local spot for a touch of dinner and some appropriate mugs of beer. The magnet was set for a long 8 hour experiment which included cycling the magnetic field up to its maximum value and then back to zero. Following dinner, Jon decided that we should probably top up the levels of liquid nitrogen and helium, so he and I went back to the laboratory. It was at this point that a mistake was made which allowed the liquid helium to blow off into the room and the equipment to shut down. It seemed to me that this had to be a result of my being in the laboratory, supposedly helping Jon with the task. But a wrong valve had been turned with the result described. Fortunately, this was not catastrophic, and the system could be refilled and restarted with only a delay in getting the resulting data. Nevertheless, another data point had been added to the discussion about theorists being dangerous, especially when they were around experimental facilities.

A Little Knowledge Is a Dangerous Thing

When I was a postdoc in Vienna, I worked closely with Helmut Heinrich on a number of different studies in semiconductors. One of these was rather unique and weird. We came across a Japanese paper that reported seeing negative magnetoresistance and current oscillations in a device made of germanium (Ge) and operated at nitrogen temperature (77 K), when one surface of the sample was sand-blasted. Normally, one expects that the application of a transverse (perpendicular to the current flow direction) magnetic field will lead to a positive increase in the resistance as a function of the amplitude of the magnetic field. This is termed the magnetoresistance. The observation of a negative value was rather unusual, particularly since this was before the discovery of the quantum Hall effect. This latter discovery initiated a great many investigations at lower temperature of the quantum properties of semiconductors. In this latter case, unusual magnetoresistance is rather common, especially as one approaches the quantum Hall region. But in the 60s, this was considered weird. So we prepared a Ge sample and sand-blasted one surface. Sure enough, we observed the negative magnetoresistance and the current oscillations. As mentioned, this observation occurred for only one direction of the magnetic field, which in itself was curious. But to investigate the underlying causes of these effects, we undertook a careful systematic study. First, we decided to orient our samples along the principle crystallographic axes, which are axes of high symmetry. To understand these, consider a Rubik's cube as a example of the basic cubic symmetry of semiconductors. The three high symmetry axes are along one edge of the cube, diagonally across the face of the cube, and the body diagonal from one corner through to the furthest other corner. It was clear to us that most transport effects can be understood by studying them for these three important symmetry directions. At the same time, we wanted to use a very soft grit (the solid material used in sand blasting). These were, of course, dangerous decisions. The effect disappeared.

The first thing we tried after this was to go back to our old grit, about which we knew nothing. Then, the effect reappeared, but weakly and only for the sample whose axis corresponded to the cube edge direction. Hmmmm! After pondering these results, it appeared that we had to learn a lot more about "grit." It turns out that it can be created with a varying range of both hardness and particle size. So, we went into the grit business. Our source must have felt like he had died and gone to heaven when we appeared and bought a dozen or so different sizes and hardness. What he didn't anticipate was that this purchase was, for us, a lifetime supply. After a great many samples, prepared with various types of "grit," we came to the realization that soft and very hard grit produced samples with no effect. A parallel effort to examine surface condition led us to conclude also that the surface preparation had no effect on the measured results. From these two sets of data, we concluded that we had a Goldilocks effect. A Goldilocks effect is one in which the perturbing force cannot be either too small or too large; it had to be just right! This is a common condition for a significant range of physical effects seen in semiconductors, always seeming to crop up in the least likely scenarios. In our case, it appeared that the soft grit was bouncing off the Ge and producing only surface damage. On the other hand, the very hard grit was eroding the surface of the Ge, removing material, and leaving only a surface effect. We needed the "just right" grit—one that did not erode the Ge, but that produced damage that went into the sample. This would yield a damaged layer some 10s of nanometers thick. This was crucial.

Some years earlier, Conyers Herring at Bell Laboratories had studied stratified semiconductors, and found that this structure would produce an anomalous positive magnetoresistance which was symmetric in the magnetic field. So, we took our measurements and separated it into two contributions to the magnetoresistance—one which was symmetric and one which was anti-symmetric. The symmetric part could now be understood by the stratification the damaged layer would produce. The anti-symmetric part was then thought to occur from the Hall effect. When the magnetic field is normal to the current direction, a force is created normal to both of these quantities. When this force pushed the carriers toward the damaged region, the resistance increased. When it was reversed to pull the carriers away from the damaged region, the resistance decreased. The culprit was the damaged region, which had a very low mobility, due to defect scattering, and thus led to an abnormally high resistance when there was no magnetic field. This was understandable, but explained the results only for the weak effect in the oriented samples. We did not understand the much larger effect that we had seen at first, and were having trouble replicating. We (Helmut and I) were of the belief that we understood transport in semiconductors, but we could not understand the results that we had obtained so far. So, it was off to discuss the problem with Prof. Seeger.

"Have you thought about the Sasaki-Shibuya effect?" The what? Neither Helmut nor I had any idea what Prof. Seeger was talking about, apparently because we were both young and ignorant. So, he patiently spent about one minute describing the effect and then sent us off to learn about it in much more depth. It turns out that almost a decade earlier, Prof. Sasaki and his students had been studying high electric field effects in multi-valley semiconductors like Ge and Si. In these semiconductors, the number of conduction electrons would be split equally among the equivalent valleys, 4 in Ge and 6 in Si. This would assure that the full symmetry would appear in measurements of things like conductance and resistance. But if you raised the electric field, the electrons would be heated by the field, and if the field were not oriented symmetric to the many valleys, the amount of heating would be different in each valley. This would lead to a transfer of electrons among the valleys which would upset the symmetry of the transport. In turn, this would lead to a transverse voltage, like the Hall voltage. But this was also a Goldilocks effect. If the electric field were made really large, then all of the electrons would be heated to such a high temperature that symmetry would return. Thus, the field had to be in the "just right" range. Moreover, the effect would disappear if the current were oriented in one of the high symmetry directions discussed above. So, our careful studies had wiped out this contribution to the negative magnetoresistance. We were off to make samples with a new orientation.

The new samples were made so that the current axis was some 25 degrees off of the cube edge direction. Voila, the large amplitude negative magnetoresistance returned. When the direction of the field was such that the Sasaki-Shibuya transverse voltage was in the same direction as the Hall voltage, the effect reached its pinnacle. And, by changing the electric field, we could further optimize the results. Moreover, we discovered that there was a negative differential conductance in the current, just as the case of GaAs discussed much earlier in this discussion. We should be able to see the current oscillations as well. And, we did. But our oscillations were not in the microwave range. We produced oscillations at the magnificent frequency of about 25 Hz. That is, these oscillations were about half the frequency of the mains in Europe. They were totally useless for any electronic device application. What a bummer. It was now clear that we would probably benefit by finding some other effect in which to invest our time. Of course, we published our paper, and in the intervening 45 years, it has been cited a grand total of 9 times! Yet, we had learned a lot, some of which may have been useful later, but I suspect not much (I did discuss the Sasaki-Shibuya effect in a subsequent textbook).

But sometimes having only a little knowledge becomes dangerous even in our teaching. A few years after the above endeavor, I volunteered to teach a course on optical propagation in inhomogeneous material, which was an interesting topic but of little interest to most semiconductors. This was during an exploratory phase of life where I was looking at a wider range of topics (including the aircraft modeling discussed earlier). If anything, this could be phrased as a difficult topic, with an advanced text.* While the course went well, in my ignorance I assigned a take-home exam, in which the presumed simple problem was to assume a very short pulse of current used to excite a frequency independent isotropic antenna at one point on the surface of the earth, and compute the time-dependent waveform at a receiving point somewhere over the horizon. The idea is that a wave at one frequency would propagate into the ionosphere, a very anisotropic and inhomogeneous dielectric medium. The higher frequencies would penetrate, but the lower

---

*M. Kline and I. W. Kay, *Electromagnetic Theory and Geometrical Optics* (John Wiley, New York, 1965).

frequencies would be reflected at some height, and could then be detected. Well, no one solved the problem immediately, so I set the group of students to work together to get a solution. The problem is that this is not the simple reflection problem that I had envisioned. Some frequencies get trapped and come out again after some distance. The atmosphere is a very dispersive medium and the dielectric function varies significantly with frequency. Thus, various frequencies encounter different attenuation and phase shifts, as well as different launch angles in order to reach the detection point. Even though one could get an analytical expression for any frequency and position of the launch and detection locations, it was still rather difficult to do the sum over the various paths. This was a problem I had not foreseen in my enthusiasm with this course. But sometimes, students overlook the difficulty of a problem and persevere. Sure enough, after a few weeks, they had solved the problem, and computed the received temporal pulse shape.

It was a significant enough calculation that we published the results of the test in a reputable journal on antennas and propagation. However, I have to admit that it received even less attention than the earlier paper on our sand-blasting experiments. Nevertheless, it had been a good learning project. Some years later, when I was with the Office of Naval Research, I would surprise a colleague, who supported work on atmospheric propagation, by knowing what conjugate points (of the atmosphere) were. It turns out that waves in the high frequency portion of the electromagnetic spectrum can get trapped in the ionosphere, where they undergo wave-guiding effects. There is a particular pair of points on the surface of the earth, one where the wave can enter the ionosphere wave-guide and the second point where the signal will come out. These two points are usually in different hemispheres. So, for example, if you wanted to listen to high frequency communications from one particular eastern European country, you would build a listening facility near Alice Springs in Australia (these days, of course, one listens to satellite communications rather than the high frequency communications that get traps in the ionosphere). Of course, other sites have different conjugate points. Like Area 51 in Nevada, these listening sites don't really exist. But if you suspect that I might be pulling your leg, you can google Pine Gap and/or Project ECHELON.

Chapter 7

The Light Side of Science

As I sit here in my cave (which is how my office is often referred to, perhaps out of consideration of that ancient samurai mentioned in the Preface) and think about the years gone by, it is clear that there have been many adventures well beyond those already listed. A number of these seemed quite humorous at the time, and perhaps will still seem to be so. Anyway, in this chapter, I will describe a number of adventures. Hopefully, they will not result in the same pain I encountered once in the Rome airport. I was obviously watching something (or someone) other than where I was going. The result was that I walked into a rather large post. It would not have been so bad, except that a colleague from the Naval Research Laboratory watched the entire episode and was now rolling on the floor laughing.

The First ASI

Our first ASI was to occur at a conference facility outside the small town of Urbino. Some events at this ASI have already been detailed, but there were more. Attendees were supposed to arrive at the train station in Pesaro, a small town on the Adriatic coast

50 Years in the Semiconductor Underground
David K. Ferry
Copyright © 2015 Pan Stanford Publishing Pte. Ltd.
ISBN 978-981-4613-34-7 (Hardcover), 978-981-4613-35-4 (eBook)
www.panstanford.com

Urbino, Italy, as it appeared in 1979.

and some 20–30 miles from Urbino. Buses were provided to carry the attendees to the conference site on Sunday, the designated arrival day. But one attendee felt the need to arrive a day late, without telling anyone about his plans. This was a young professor from one of the Texas schools, whom I will call Dr. S. On Monday, we received a phone call that he was in the train station in Pesaro and wanted to know when we would pick him up. Well, we didn't plan to, but finally arranged for a car to bring him from the train station. In an earlier time, Dr. S had acquired a nickname as the "little boy." This had occurred at a major international conference in Rome. The conference excursion had gone to the ruins at Ostia Antica. The organizers had carefully arranged that no one would be left there, and so people were asked to carefully remember on which bus they had arrived. Dr. S had come on our bus with a great many of our shared colleagues. When it came time to return to Rome, our bus could not leave, as we were short one person. When the driver asked if anyone knew who was missing, one person said that they thought it was a little boy who had been on the bus. So, the driver went and searched, finding nothing. Presently, someone got off another bus, which apparently also had not left as it had too many people. Here came Dr. S, in his shorts over to our bus. As he climbed aboard, the driver remarked "Here is the little boy." This name has stuck with him ever since.

Anyway, as the discussion in Urbino proceeded about how to bring him from Pesaro, it appeared that his nickname became known to a rather wider audience than just the attendees. After the events described earlier, we finally arrived at the dinner on the last night at the conference facility. After dinner, several of us migrated to the bar. The bar tender was a regular who allegedly did not speak English. From many encounters, however, I doubted this and thought that it was probably just a defense mechanism. Anyway, Dr. S joined us at the bar. I had ordered a Gran Marnier. Dr. S asked what I was drinking and whether or not it was good. I assured him it was, so he ordered one. Just as he was about to take his first drink, I said, "But, it is about 1000 calories!"

Karl Hess and Hal Grubin, during an investigation of the town of Urbino.

Dr. S put down the Gran Marnier and ordered a Coke. The bartender brought him his coke and set it down. Then he slid the glass of Gran Marnier over to me, stating (in Italian) "For the director, Gran Marnier. For the little boy, Coca-Cola." At this point the bar erupted, as all but Dr. S had figured out immediately what had been said.

The laughter brought Prof. J to the bar and he joined the festivities. During the course of the ASI, Prof. J had been leading groups of attendees through the fields as a shortcut to reach Urbino, so that they could enjoy the town's night life. It probably didn't hurt that there was a music school in town, and this school attracted a significant number of young English women. Anyway, knowing that Prof. J had a deathly fear of snakes, I asked him if he had read the signs outside the center at any point where one might enter the fields. He replied, "No, they are all in Italian! I can't read those." So, I told him that they warned everyone that the fields were infested with poisonous vipers. His eyes got big, and I don't think he slept well that night.

Some years later, Prof. J visited Arizona State. In the old life sciences building, one hallway had a series of windows which opened onto displays of the various snakes and critters which

A confused conference committee wondering what they were supposed to do at this point.

could be found in Arizona. So, on his visit I took him to see this display. He apparently thought that all of the various reptiles had passed through the hands of a taxidermist, as he bent over and carefully peered into the windows. But, as he did so, one snake which was laying very close to the window raised its head to peer at him. At this point, I basically had to peel him off the ceiling. And, he definitely had nightmares that night.

Just Deserts

We were at a large conference in Montpellier a couple of years later. Of course, Prof. J and a number of close collaborators were also there. During this conference, Prof. J had gotten on the nerves of a number of speakers, which in a sense was his normal behavior. One evening, there was a scheduled dinner for the conference committee, which included the two of us. We boarded a bus that was to take us to the dinner at a nice seaside restaurant. I suppose the organizers wanted to entertain us, and so they arranged to have the bus stop at a site where we could enjoy the ocean. The problem was that where the bus stopped there was no beach. Between the pull-off and the ocean was only a pile of large rocks. We couldn't change and swim, and it was very difficult to climb down to the sea. So, mostly, the committee members just kind of stood around in confusion. Finally, however, we were encouraged to board the bus once again, and proceed to the restaurant. Our table was well organized, but was parallel to the beach. Prof. J complained about having his back to the ocean, but I told him, "It was obvious when we came in that only half of us would have an ocean view. Why did he act so slowly to select his chair?"

As the evening wore on, he gradually managed to get on a great many people's nerves once again. Finally, there was a spontaneous eruption, in which several people jumped to their feet (I have to admit being one of these). We grabbed Prof. J's chair, with him still sitting in it, and carried it to the surf. These two objects were then deposited in the surf. There he was left, with the chair on its back, and his feet in the air, waves lapping at his body. There then was a cheer, and we went back to eating and drinking. Because of the event, more wine was brought to the eager crowd. Prof. J said later that he saw his life pass before his eyes as he was being carried away, and was certain of his doom. In any case, his behavior was well moderated for the remainder of the conference.

San Miniato

Our second ASI came in 1983, a couple of years after the Montpellier incident. This was held at a conference facility a few miles outside of the small town of San Miniato. The town lay on the old highway about half way between Florence and Pisa. The facility was a former cloister which had been acquired by a bank and recast as a conference venue. Nevertheless, the main lecture hall was the old church and one had a somewhat larger encouragement to be truthful in his lectures. Even though the

Carl Wilmsen, my experimental colleague from Colorado State, having lunch in Pisa during the weekend outing.

venue was a few miles from town, this did not stop the attendees from drifting into town for the night life, and to attend a music festival that was occurring in town.

At one point, I was giving a lecture on the interactions between devices, and how this could lead to quantum like behavior. A well-known condensed matter theorist from Hoboken asked if I could give an example of some Hamiltonians for the process. At this exact moment, a colleague of mine from Colorado State walked in. Now, Carl Wilmsen was an experimentalist, and the question brought him up short, stopping him in his tracks. He leaned down to a nearby attendee, saying "What did he ask?"

The attendee responded that he had asked for some examples of Hamiltonians. Ever the experimentalist, my colleague responded "I knew I was in the wrong session," and immediately walked out of the room. The problem was that we did not have parallel sessions.

On the last morning, I was sitting and eating breakfast. The previous night, we had a gala banquet, attended by all and featuring copious amounts of refreshments. On this day, we had only a half day's worth of lectures, with everyone departing after lunch. Well, on this morning, Dr. M came wondering into the breakfast room, and he looked terrible—unshaven and his hair uncombed. This was completely out of character. Dr. M was a

Bruce McCombe and Norman Horing resting during the excursion in Pisa.

scientist from the Naval Research Laboratory, and was to deliver the first lecture of the morning. As he sat down, he remarked that he apparently had lost his toiletry kit. He had packed the previous evening, so had to unpack everything but still couldn't find it. This accounted for his appearance. Shortly after this, an attendee—a young woman from what was then the Royal Radar Establishment in Malvern—came into the room with an obvious injured ankle. Hmmm. This started to sound like there was a story behind these occurrences. As time passed, the story obviously came out.

After the gala banquet, a number of the attendees had gathered in a room on the top floor in order to sample and share a reasonable amount of single malt scotch, which he had brought to the ASI. As the evening wore on, they discovered a door that led to the roof of the building. Since these were scientists, it was only natural that they explore the roof (this was several stories up, I am told). In their wanderings, they came upon an open window to a bathroom for one of the lecturer's rooms. This was too good to pass up, and so a few of them plunged through the window. Among these was our young colleague from Malvern who rather fell through the window, giving rise to a severely sprained ankle. The intruders gathered up the toiletries and left. This, of course, left Dr. M in the lurch the next morning, at least until the toiletries were discovered on the overhead projector at the start of this lecture. I suppose it was fortunate that he had closed the door to the bathroom, or other things may have been relocated.

One of the other lecturers was Karl Hess, from Illinois. As fate would have it, he was at a small conference in London a few weeks later. At one of the meals, he was regaling the attendees with the story of San Miniato. As he was getting into the story, he noticed a young woman at the table who was starting to blush. It was our young woman from Malvern. So, the story took on a more personal tone, and became much more interesting to the other people at the table.

The Duel

In the 1990s, we held two ASIs at Il Ciocco, a conference hotel some 20 miles north of Lucca and very close to the small town of Barga. As it turns out, this was on the old road over the mountains from Modena, which made this old road of interest to test drivers from Ferrari, but to no one else in their right mind. Il Ciocco was a great site for our conferences because the room cost included the meals and all the wine we could drink at the meals. At the first of these ASIs, I estimated that our group was consuming about a bottle of wine per person—*at lunch*. When, we returned for the second time, my co-organizer Carlo Jacoboni was talking with the general manager one morning. The GM remarked that they had not had a group drink so much since . . ., then his eyes got big and he said to Carlo, "That was you guys!" So, we had apparently earned a big reputation in at least one category.

As usual, the small town of Barga provided a location for the evening excursions by the attendees, and it became known for its food and wine. Now, Il Ciocco was known also as a training center for some soccer teams as it had a sizable stadium at the top of the hill upon which it was located. Of course, we couldn't use this stadium, but there was a small field nestled nearby (that we called the quantum field). So, we organized a soccer tournament,

The author and his wife (left) with colleagues during an outing to Florence at the 1990 Il Ciocco ASI.

Steve Goodnick, Carlo Jacoboni, and John Barker (from the left) enjoying a lighter moment during the 1994 Il Ciocco ASI.

and managed to get four teams together during the first of the ASIs. A special dinner and wine was organized for the winning team, that was not our team of lecturers. Our team suffered the normal excessive zeal of people somewhat older than the young attendees. As a result, one of our lecturers had to leave to arrange treatment for cracked ribs. Our team suffered the indignity of one and done! We lost our first game 4–1.5. You may be wondering how we scored a half goal. The goals themselves had no netting. Each goal post was supported by a tube angled to the back. Thus, one elegant shot went between the post and its support on both sides of the goal (parallel to the goal line). The referee was so impressed that he awarded us a half point. The referee may have been biased, as he was one of the lecturers as well. However, the other team was not bothered as they knew they were going to win anyway.

At the second ASI, there occurred an event that has become legend, at least to some of us. Carlo organized a gala evening dinner just for the lecturers and organizers. Carlo has been a valued friend and colleague for more than four decades now. Yet, on this evening, he and I had a small dispute. It was decided that we would have to settle this with a duel. Our weapons of choice were bottles of spumante. So, after the main course, but before

The duelists are prepared.

the dessert, space was cleared away for the event. There we stood, back to back, with our bottles raised. The umpire gave a sign and we took a few paces away from each other before turning. As I turned, his cork flew past my knee and bounced on the floor. I had him! I leveled my bottle and carefully launched the cork. It struck him in the middle of his chest. Victory! He complained that I had shaken my bottle during the paces, which he asserted was illegal. First, how did he know this if he was supposed to be facing away from me, and second, no one laid down such a rule. It was finally decided that we would just have to drink the spumante.

This was to be our last ASI, for reasons that I have already expressed. In fact, I think that the overall numbers of ASIs has gone down over the years since the cold war ended. Perhaps that is as it should be, but it is just another loss for the spread of scientific knowledge among colleagues around the world.

The Chateau

Among the ASI's discussed above, I was invited to lecture at an ASI held in France. This particular ASI was organized by colleagues from Nottingham and Toulouse, and was held at a chateau west of Toulouse. I would have to say that this experience surely reinforced my preferences for meetings in Italy.

The chateau was a nice building, but only the organizers and those with their wives present were housed in it. For attendees and lecturers, there were other arrangements. One of these was a set of rooms that were carved into the side of a hill, but with reasonable front sides. When registering, participants were issued with a relatively large metal container about the size of one of those fire extinguishers that hang on a wall. There was, of course, a hose from the bottom, and a large handle was present at the top of the tank. It was explained to the attendees that, as the chateau was a working farm, there was a small problem with bugs and insects. The tank contained an insecticide with which the roommates were to do battle with these bugs and insects. One of my colleagues from ASU, who joined me at the ASI, remarked that they were apparently losing the war as they had to have their tank refilled a couple of times during the two week stay.

I was apparently more fortunate, as my assigned room was in a building separate from the hill. My roomie was a scientist from the Naval Research Laboratory and our room was on the second floor, over a set of rooms that had been converted into offices, one of which was used by the organizers. We were issued only a small, standard size can of bug spray that we used extensively upon arrival in the room and then only once

Two attendees congregated outside the office area of the ASI at the Chateau.

more during our stay. This kept the population of bugs and insects to an acceptably low level. It turns out, however, that our room was very conveniently located, as it became apparent shortly after arrival that the organizers had a stash of single malt scotch somewhere on the premises below. For two such good scientists as my roomie and myself, it did not take long to ascertain just where the hiding place was located, and it was within the office area below our room. Thus, we spent several enjoyable evenings on the patio outside the office sampling the wares brought along by the organizers, sometimes with them present.

The lecture hall was also burrowed into the side of the hill, and had apparently been a wine cave at one time. It was only wide enough for perhaps 10 chairs abreast, and had a single entry at the back, although there were several windows along the back wall (the lecturing end was under the hill at the end of the cave. There was not much heating as this was France in summer! But, within a few days, the lecture hall was referred to as the incubator, and in a not particularly humorous manner. It seemed that the number of colds and bad allergic reactions was growing exponentially among the attendees. By the second week, more than half of the latter had moved the chairs outside the hall and watched through the open windows of the back wall.

Now, this chateau was located about 15 km from the small French town of Condom. My colleague from ASU was a postdoc in the group and was particularly entranced by the existence of this neighboring town. He became obsessed with buying one of those accoutrements which bear the same name in America (he was not aware that it had far different names in Europe, particularly in France). So, one afternoon, he rented a bicycle from the chateau and set out to see Condom. Of course, he spoke not a word of French, so finding the right shop and explaining what he wanted was going to be a problem for him. As the hours passed, and we went to dinner, it became known to all of the attendees that he was on a quest. So, his return was eagerly awaited. Just as dinner was ending, he made his entrance, and the clamor began to know if he had been successful. It apparently took him a couple of hours in the town to find the right shop and to describe what he desired to buy. But he was successful. However, on the return trip, the bicycle had developed a flat and he had been forced to carry it the last few miles back to the chateau. Nevertheless, I am sure that he planned to tell his grandchildren about his great quest to find a condom in Condom.

The Challenge

At some point in preparing an outline for this book, I apparently had a brilliant idea for this section, but I have forgotten what that idea was. Now, I have been able to forget things throughout my entire life, and I don't believe I am much different than anyone else in that regard. But it seems that with passing years, my forgetfulness has increased. I know that this is a trait that is assigned to people as they get older, so apparently it is a widespread problem, or so people say. My long time friend and colleague Karl Hess says that we are afflicted with CRS syndrome (which, so I am told, stands for "Can't Remember S*&^"). I don't know whether Karl learned this from others, or if it is original, but apparently he is also afflicted with this problem (or so our wives tell us). Such a trait can be useful, for example, when my wife complains about my driving too fast, I just have to tell her "I have to get to where I am going before I forget where it was that I was going." I admit that forms of this phrase exist on the web, but that doesn't undermine its usefulness in real situations.

One of the most common occurrences of the problem arises when, perhaps after wandering a bit, I enter a room where upon the question "What am I doing here?" enters my mind. Now, this is not a deep philosophical problem where one wonders why they exist. I don't worry about the deep philosophical question, as I have come to grips with it—I am a professor, and hence I am here to warp young minds, and to perhaps turn these youngsters into productive engineers and scientists. It is far too difficult to try to determine any deeper purpose, and probably also far too frustrating to spend time on such questions. The much less esoteric problem of arriving in a room, without any understanding of why you arrived is more perplexing. This is because of the chance that your state of confusion will be observed and interpreted as another sign of the onset of senility. It is not that at all. You must remember that the public perception of a scientist is someone in a white lab coat, with unruly hair, and a natural predilection to be forgetful. It is assumed that forgetfulness is a result of deep thinking on an important topic, so that the environmental surrounds are pushed out of our envelope of awareness. So, arriving at the wrong room is easily explained—I was in such deep thought, I lost contact with where I was going.

I once had a colleague to which this deep thinking process often occurred while he was driving. Unfortunately, when this happened, he proceeded directly down the street in a reasonably straight line, but did so without any awareness of traffic lights. This was not good! Fortuitously, I have been able (so far) to avoid such states while driving. Oh well, it is known that getting old is not for the faint of heart.

Fame Comes in Strange Packages

Klaus von Klitzing was young when carrying out experiments at the high magnetic field facility in Grenoble. There, he made the experiments that discovered the quantum Hall effect. A few years later, he was awarded the Nobel Prize for this work. Of course, this led to great fame beyond just the scientific community. But fame comes in other ways. For Klaus, it was a wine named for him. Bob Laughlin, the 1998 Nobel Prize winner for his part in the fractional quantum Hall effect, says that he became aware of

The Ferry bar.

it on a cruise down a river (while visiting Klaus in Germany) as they passed the winery.* On the hillside, students were holding a sign for the von Klitzing wine. The label is covered with symbols describing the quantum Hall effect. I have actually had some of this wine in recent years, and it is not bad.

In the Fall of 2013, the Institute of Physics published a new book of mine, which is titled Semiconductors: Bonds and Bands. This is both an ebook and the normal hardcover variety. Now, the IOP always holds an evening reception at the annual March meeting of the American Physical Society. The year after the book was released they invited me to come to the reception and talk about my experiences in writing an ebook, which is somewhat more troublesome than a normal ebook due to the equations and figures that need to be included. So, I accepted the invitation and went to Denver at the appointed time. At conventions such as this, they always hand out trinkets at the door as people come to the reception. For this occasion, the IOP had had candy bars made with the book's cover on the wrapper of the candy bar. Here I was, with my name on a candy bar. Klaus has his wine, but I have a candy bar. I suspect, however, that his wine will be around much longer.

---

*R. B. Laughlin, *A Different Universe* (Basic Books, 2008).

The Proper Response

The one task that all scientists face is having to deal with critics. In particular, reviewers of scientific papers seem to have a feeling of superiority that gives them the right to be uninformed and rude. Now, don't get me wrong. Usually, the reviewers are conscientious and do a pretty good job, and in some cases they even make suggestions which improve the paper under consideration. We submitted one manuscript to *Physical Review Letters*, a preeminent journal of the American Physical Society. One reviewer in particular gave us a review that was as long as our manuscript, but it was particularly insightful. We completely rewrote the paper and went for another round of reviews. Again, this reviewer came back with insightful suggestions which led to a further rewriting of the paper (one thought was that he should have become a co-author). Finally, his last comment was that the paper was now too good for this journal! Nevertheless, the journal published the paper without further comments.

Let me return to the opening comment. There is a small fraction of the reviewers who take the anonymity of the peer review process as an opportunity to engage in dictatorial politics. Here, they view the review process as an opening to challenge the work on a variety of levels, most of which are irrelevant or unrelated to the work itself. One colleague complained about the reviewers comments on his paper. When he replied that the reviewers didn't know what they were talking about, the editors subsequently rejected the paper instead of seeking an independent comment from a different reviewer. This happens a lot when one challenges the status quo. To be honest, editors have to handle a rather large number of papers at any good scientific journal, and sometimes they just don't have time to be incisive in each case. However, I have had an editor go through 5 or 6 reviews before coming to a decision. In one particular case, the recommendations were always 2–2, or 3–2, or 3–3, which should have told the editor that the topic of the paper was a little controversial. In the particular case, the editor came to this conclusion and accepted the paper, which became a highly cited paper.

Sometimes the reviewers really miss the point. One paper in particular arose when I was with the *Journal of Physics Condensed Matter*. For this paper, two reviewers had rejected the paper, but for the wrong reasons. The author came back with a long letter to the editor (me) expounding upon the importance of this work. So, I read the manuscript myself along with the reports of the two referees. The problem was that the author had no introduction. Instead he immediately started the manuscript with the first of a long number of equations. The work was interesting and should have been useful to a particularly active area of current research. But the way the manuscript was put together made it a barrier to any reader. I suggested that he rewrite the paper, and particularly put in an introduction with all the material and rationale that was in his letter. After rewriting the paper as suggested and adding more discussion for the reader, the paper passed the review process with flying colors and was published. It garnered almost a 1000 downloads in the first six months after publication and wound up being among the outstanding papers for that year. The first reviewers had merely taken the opportunity to point out that the paper was boring and not likely to be important, a point that they got wrong. Worse, they didn't communicate how it might be improved, so the reviews were not useful. But they had done their job!

One of my colleagues took a different tack. For one of his manuscripts, the single review came back and was entirely useless. The reviewer apparently did not understand the work and was not familiar with its applications. It was relatively well known among our group at the time that I had acquired a particular rubber stamp during a visit to San Francisco. This stamp was "bulls*&^" in all caps with 1 inch high letters. My colleague's students thought that he should be better endowed, so they bought him the equivalent rubber stamp, but in 2 inch high letters so that it just fit diagonally across a standard page.

My colleague figured he had nothing to lose, so he printed out the reviewer's comments and applied the rubber stamp to them. Then he sent the page to the editor of the journal, and asked him to forward it to the reviewer. We don't know if the editor sent it to the reviewer, but he did take the hint and sought out another review of the paper. This time the paper was accepted.

Chapter 8

Arrogance and Ignorance

Sometimes you hear a phrase that just sounds so correct that you immediately think of just where that phrase should be applied. Such is the case for a phrase that originated with the president of my university. Michael Crow has led Arizona State University for more than a decade now, and he is known quite widely for the directions he has taken the university. He is molding us into what he calls the "New American University." Along the way, we have received a great deal of new publicity and recognition, and both the physical plant and the student body has grown accordingly. But some of his ideas are at odds with those of many other leaders, some of whom have beliefs still stuck at those of a university of some years past. Crow was giving a talk at a morning meeting of leaders from the west side of the Phoenix metroplex. In the talk, at least as it was reported in the newspapers, he was addressing the principles of the "New American University." Along the way, he commented that some of his east coast colleagues disagreed with his views, and then noted that some of these easterners were best characterized as "arrogantly ignorant." Wow. We always knew that he could be

50 Years in the Semiconductor Underground
David K. Ferry
Copyright © 2015 Pan Stanford Publishing Pte. Ltd.
ISBN 978-981-4613-34-7 (Hardcover), 978-981-4613-35-4 (eBook)
www.panstanford.com

blunt and plain speaking, but when reading the press report, I immediately knew exactly what he was talking about (as may be seen in the title of this chapter). It was clear that I had to adapt the phrase to describe a number of items, some of which are discussed in this chapter. The phrase describes an entire mindset in a succinct and direct manner, but I have modified it a little bit to be more descriptive of a broader range of thinking.

Sometimes the Theories Are Wrong

At some point in the mid-'80s, I was attending a conference in Santa Fe, NM. The conference was devoted to the physics of semiconductor interfaces, and I was in a session concerning the various theories of interfacial structure. This session was chaired by Charley Duke (at that time a scientist with Xerox's research center in Rochester, NY). A young Asian scientist was delivering her talk on this topic, and she was taking a lot of heat from the audience. She just wasn't handling the questions very well, when Charley decided that the fairest thing to do was just cut off the question period. As he did so, he remarked "All theory is wrong, it is just a question of degree." This became the origin of what I now call Duke's law. One of my research associates, Wolf Porod (now a chaired professor and center director at Notre Dame University), provided the corollary: "All experiments are correct, but what did they measure?" These two statements are so straightforward and clear, with very deep insinuations, that it continues to surprise me how otherwise competent scientists can fail to appreciate their import. We find these same scientists pontificating upon their results without ever considering that they are subject to these simple rules. If one doesn't understand and appreciate the limitations placed upon any results by the approximations on one hand and the experimental apparatus on the other, then it is difficult to ascertain just how believable the results can be.

When semiconductors have a free surface, the forces on the individual atoms are different than those within the interior of the material. As a result, these surface atoms can move around to find a different energy minimum. If the atoms move without changing the basic unit cell, it is called relaxation. If, on the other hand, the surface unit cell is different than that of the bulk, the surface is said to have reconstructed. The new, surface unit cell is usually described by a symbol such as 2×1, or so on, where the units are in terms of the size of the bulk unit cell. For many years, the theorists had done calculations to predict which reconstructions were most likely to occur, and which conditions favored one or another structure. Surface experimentalists, working in high vacuum, usually cleave the crystal *in situ* so that they can study the pristine surface that remains after cleavage. With silicon, this is the (111) surface (the surface that is normal to a body diagonal of the cube). Some experimentalists thought their surface sensitive probes indicated that there might be a 7×7 reconstruction of this surface, and many theorists supported this. But the main body of computational theorists said this was unlikely. This is a very complicated reconstruction, and the latter theorists held sway among the community until 1983. Gerd Binnig and Heinrich Rohrer invented the scanning tunneling microscope (STM) in 1981 (for which they received the Nobel Prize), and investigated the silicon (111) surface among their first studies. The STM works by bringing a very small, sharp metallic tip near the surface of a conducting material, so that electrons can tunnel between the tip and the surface. By moving the tip in a manner that keeps the tunneling current constant, they can monitor the very small variations in surface height with atomic precision. This means that they can map the surface topography with this same atomic resolution and provide a real space image of the surface. The result of their study on the silicon surface was that the 7×7 reconstruction seemed to be the dominant structure.[*] So why did many theorists object to this reconstruction. It seems to me that the problem lay with the computers of the day. The

---

[*]G. Binnig, H. Rohrer, Ch. Gerber, and E. Weibel, *Phys. Rev. Lett.* **50**, 120 (1983).

complexities of this reconstruction require quite a large number of atoms to be included in the computation, and there were very few places where the required computing power was available. So, if they couldn't compute the properties of the structure, then it must not exist! This view is not unlike the philosophy discussed in an earlier chapter, where Mach did not believe in atoms because he couldn't see them.

When scientists started making very small electronic structures, they discovered a variety of new effects. Since the first such small structures were made from metals, such as metal wires, the size was not all that small when compared, for example, to the particle wave length at the appropriate energy within the metal. So, these new effects were given the name mesoscopic phenomena.* When the effects were also seen in semiconductors, it was clear that many of them arose when the mean free path (between electron scattering events) was comparable to the size of the structure. But many of the properties were quickly attributed to the presence of disorder within the material. In semiconductors, it is presumed that the disorder arises from the impurity atoms in the device. Each ionized impurity, such as a donor which is ionized by exciting its outer electron into the conduction band to contribute to the conductivity, has a local potential which modifies the overall potential landscape within the device. Normally, the impurities act independently to provide only weak scattering rather than a larger, somewhat coherent behavior leading to the observed mesoscopic processes. But with a large density of impurities at low temperatures, such coherent processes are more likely. One such process, and the one in which we are currently interested, is the occurrence of conductance fluctuations. Now, one usually thinks about conductance fluctuations as being time dependent events that arise from the random variation of the electron density, and are generally referred to as *noise*. Such noise has been studied for a great many

---

*See, e.g., C. W. J. Beenakker and H. van Houten, *Sol. State Phys.* **44**, 1 (1991).

years. Here, however, the conductance fluctuations are not time varying, but are seen as one varies the Fermi energy or applies a magnetic field. These fluctuations generally are repeatable (when, for example, sweeping a magnetic field through a range of values several different times). It was assumed that these fluctuations were a result of quantum interference in the disordered potential landscape (which means the spatial variations of the local potential in the device). One can picture this by considering a range of mountains spread over a particular area. As the great flood approaches, the water rises first in the valleys between the mountains. This water can be thought of as the sea of electrons in our semiconductor device. When the level of the water is close to several of the passes through the mountains, a small change in the depth will dramatically change the layout of the observable mountains (and the water). So, the actual path which a boat follows in navigating through the mountains can vary dramatically with small changes in the depth of the water. The number of paths which the boat can follow corresponds to our conductance through the device. The fluctuations of conductance correspond to the fluctuations in the number of paths the boat may follow. When the mountains are reduced to the quantum mechanical size scale, then two paths which flow around the same peak can interfere just like the two-slit experiment. As the depth of the water varies, the interference can lead to peaks and valleys of the conductance as the interference moves from constructive to destructive interference.

A group of theorists carried out a difficult quantum mechanical calculation to determine the expected amplitude and scale of the fluctuations that should be observed. Of course, they obtained results which were comparable to that seen in the experiments. But then, they postulated that these fluctuations were universal in nature. That is, they stated that these fluctuations would have amplitudes (of the conductance change) which were independent of the amplitude of the disorder. They also stated that the fluctuations should be ergodic; they would have a well defined relation between the observations found for a variation of the Fermi energy and a variation in magnetic field. Thus, from one calculation, they jumped to a grand unified theory of the conductance fluctuations. Because of its grand scale, this theory was now accepted almost without question by the community. The problem, of course, was that such a theory was counter-intuitive. In our mountain example above, if I use a plow to level out all the mountains, I can get a uniform depth of the water and all of the interference effects should go away. Indeed, the GaAs/AlGaAs heterostructure discussed in an early chapter could produce the quasi-two-dimensional electron gas with exceedingly high values of the mobility (which means very little scattering). In fact, devices made from such material exhibited essentially zero conductance fluctuations.

This question about universality was of interest to our group for many years, which was natural as we pursued our investigations into ultra-small semiconductor devices. As companies like Intel pushed down the size of individual transistors, the presence of such effects would greatly affect their performance and could mean an end to Moore's law. But this didn't seem to be occurring. So, we spent some time studying the role of the amplitude of the disorder. Sure enough, we found that the fluctuations went away as the amplitude of the disorder was made very small (or zero).[*] The fluctuation amplitude went up linearly with that of the disorder until a plateau was reached. But the conductance of the entire structure decreased as a whole as you raised the disorder.[†] Hence, there was no universality with respect to the amplitude of the disorder. We also discovered that the variation with magnetic field produced a much smaller fluctuation than predicted by the detailed theory. Moreover, we found that one did not really need the detailed theory to predict the amplitude of the disorder. Many years before, Rolf Landauer (mentioned earlier) had shown that, when the size of the device was quite

---

[*] A. Grincwajg, G. Edwards, and D. K. Ferry, *Phys. B* **218**, 92 (1996).
[†] B. Liu, R. Akis, and D. K. Ferry, *J. Phys. Cond. Matt.* **25**, 395802 (2013).

small, quantum effects would create transverse modes just like in a microwave waveguide (the modes are like the modes on a string instrument).* When the plateau of the fluctuation amplitude was reached, it was found that this amplitude was given by turning off and on a single one of these transverse modes. So, the Landauer theory gave us the amplitude of the conductance fluctuation without any need for the more complicated theory.

Early Quantum Mechanics

Earlier, in Chapter 4, we discussed the early philosophy of Ernst Mach and its adoption by Neils Bohr. This philosophy creates a number of problems with the interpretations of quantum mechanics and has kept the philosophers quite busy over the years since Bohr developed his model of the atom. Bohr introduced his model of the atom in 1913, with the idea of a central nucleus surrounded by the electrons organized into a series of concentric shells. Following a number of different models of the atom, Rutherford had suggested in 1911 that the nucleus should be a small heavy core with the electrons circling around it. But it seems to be Bohr's idea that the electrons were organized into shells. The problem was that he had no idea why the electrons formed into these shells. Nevertheless, he postulated that these shells were stable and would not decay as one expected from classical physics. This was the revolutionary idea, but one whose root physics was not understood. On the upside, the results of Bohr's calculations for these shells were confirmed by the known experimental data on absorption and emission of radiation by the atom. Yet, it came to be known, especially by later theoretical work of the new quantum mechanics that Bohr's calculated energy levels were wrong in detail, even if their spacing was correct (which was important for the experiments). Perhaps the lack of an underlying physical understanding of the rationale for the shells led Bohr to the adoption of Mach's philosophy. The shell

---

*R. Landauer, *IBM J. Res. Dev.* 1, 223 (1957).

model of the atom could not be understood from basic physics, but could only be determined by direct experiments. Reality lay only in the measurements, not in any underlying causative effects. Strange as it may seem, such a belief still carries over to today, where it is often explained that quantum mechanics can only give you possibilities and probabilities, reality arises from the measurements. Nevertheless, Bohr's theory was considered to be earth-shaking, even though it was incomplete, in that there was no known reason for the electron shells to be stable.

By the time of Bohr, it was already known from Planck's work that light could behave as particles, and this led to Einstein's work on the photoelectric effect which earned him his Nobel Prize. It was not at all clear that this was a two way street, and that electrons could also behave as waves. It was up to Louis de Broglie to make this connection. In his doctoral thesis in Paris in 1924, he suggested that particles could show wavelike behavior, and more importantly, this would lead to the shell behavior in atoms. The story goes that his advisors weren't sure that the thesis was correct, and only after Einstein read it and suggested that it could be right, was the degree awarded. To explain Bohr's shells, de Broglie suggested that the energy of the particle was connected to a wavelength for the electron in a similar manner as for the photon (although with a different relationship, now known as the de Broglie wavelength). De Broglie's important recognition was that the circumference of each shell had to match to the wavelength of a particle with that energy through this relationship. If the energy were increased slightly, this would lead to a smaller wavelength, which would counteract this increase. Thus, only particular connections between the wavelength and the energy were allowed. This provided the underlying physics which explained Bohr's shells. Erwin Schrödinger used these wave ideas to then development his wave equation, whose solutions provided the correct energy levels for the atom. This was one of the foundations of the new quantum theory, the second came from Heisenberg and his particle-based matrix mechanics formulation.

But it appears that Bohr did not completely accept this idea of a wave basis for the particle. I guess to do so would have required him to admit that he missed the important physics in his atomic model and that perhaps his philosophical interpretation was to be challenged. If one could calculate all the correct energy levels (the shells) and also the transitions between them, then the experiment only confirmed what was already known. Reality existed before the experimental measurement. The story of Bohr inviting Schrödinger to Copenhagen, and then brow-beating him to such an extent that he became physically ill, is well known. But Bohr could not induce him to give up the wave theory. Bohr carried on with the development of a notion of wave–particle duality and his theory of complementarity. In a simple form, wave–particle duality meant that the two approaches (waves and the matrix theory) were equivalent and should give the same answers. Complementarity asserts that the measurement must give the classical results, but can vary with the apparatus used in the experiment. These two assertions have come be used as the basic tenets of what is called the *Copenhagen interpretation*. But there are very good reasons to believe that Bohr was wrong in these assertions.

It was only a short time later (1927) that the experiments were done to demonstrate the wave nature of the electron. George Thomson and Clinton Davisson shared the prize for their separate demonstrations of this wave behavior. In both cases, the diffraction of electrons from a crystal of a metal was used to demonstrate the wave nature. This is important because our semiconductor physics is based upon just this effect. Our understanding of the energy bands, effective mass, and scattering processes are all defined by the wave nature of the electron. Hence, the way in which quantum mechanics is understood to be, and the manner in which its philosophical foundations are established, are quite important to our understanding of semiconductor physics. It is probably worth pointing out that every textbook with which I am familiar develops these theories from the wave version of quantum mechanics—the Schrödinger equation. The observation of electron waves became easier with the development of the STM, discussed above. In studying metal and semiconductor surfaces, one could observe atomic steps along the surface, and also see the remnants of wave behavior of electrons localized near the free surfaces. Don Eigler, and his colleagues from IBM's Almaden research center, found that they could position individual atoms on a metallic surface using the STM, and created a quantum corral—a ring of atoms with an electron trapped in the center of the ring.* Clearly observable in the STM images of the structure was the magnitude of the wave function for the electron captured in a cylindrical potential well. The remarkable fact was that one could now see the wave function—it was a real quantity that could be observed. Eigler was known to have strong opinions about Bohr's wave–particle duality, when he stated:

> "I don't believe in this wave–particle duality mumbo-jumbo. I think it is mostly just the left-over baggage of having started off understanding the world in terms of particles and then being forced, because of the quantum revolution, to think of the world in terms of waves, and we are stuck with this dualistic way of looking at these very small particles. Don't even think about them as particles, electrons are waves. And, if you think of them in terms of waves, you will always end up with the right answer."[†]

This is a pretty powerful statement about the nature of quantum mechanics, and it is one that I can fully understand and with which I agree. Much of the confusion that surrounds various interpretations of quantum mechanics goes away if you stick with this single-minded dependence upon the Schrödinger wave mechanics.

---

*M.F. Crommie, C.P. Lutz, and D.M. Eigler, *Science* **262**, 218–220 (1993).
[†]http://www.youtube.com/watch?v=JKatDurrDHM. Eigler's image of the corral and his comments begin at about the 7:35 mark of this video.

But the struggle against waves, in spite of Bohr's wave–particle duality, led to a great many arguments over the years. Perhaps the most famous is the set of debates and arguments between Bohr and Einstein. Einstein was a realist, and accepted that the wave function provided such a real quantity, even if it were only related to the probability of finding a localized particle. Bohr, as we have noted earlier, felt that the reality didn't exist until the measurement of the particle had occurred, since it was then projected into the classical world. The argument led to some consequences, since the measurements yielded values for only a few variables (or parameters). If there was a reality prior to the measurement, the dominant view was that there had to exist other variables (or parameters). Since these weren't measured, they must have been hidden. Such hidden variables were not acceptable in the Bohr model of the world, since they were not measurable. Roughly from this time, ideas for proving that quantum mechanics was correct meant that one had to prove that hidden variables could not exist, and that Bohr's interpretation was the correct one.

In fact, all of the early attempts to prove that hidden variables could not exist eventually failed as errors were found in their logical basis. The one that has lasted the longest is Bell's theorem.* Bell tried to show that if one had hidden variables, the assertions of reality would lead to an inequality that had to be satisfied. Here, he assumed that the condition of reality would mean that two variables were statistically independent (no correlation could exist between them). Consequently, experiments designed to test this inequality with quantum systems proved that these quantum measurements violated this inequality, and this led to the assertion that quantum mechanics was correct, and there could be no hidden variables. The problem with this approach was the fact that Bell's inequality was already known for almost a century, which means that it predated quantum mechanics! Hence, it is likely that it also affects classical mechanics, and its

---

*J. S. Bell, *Physics* **1**, 195 (1964).

violation has nothing to do with quantum mechanics. In fact, an example of patients with a particular disease has been given, in which different methods of gathering data can either obey or violate the inequality.[*] Other violations have appeared in classical fluid dynamics[†] and in the behavior of coupled pendula.[‡] The key to this is that the presence of any correlation can lead to a violation of the inequality by rather trivial methods. The impact of this is that the so-called proofs of quantum mechanics, which rely upon violations of the Bell inequality, are not proofs at all. In particular, most of these measurements cannot distinguish between quantum entanglement and classical correlation.

Quantum Jumps

In the previous section, we discussed how Bohr generated a model of the atomic structure in which he postulated that electrons moving around the nucleus would have stable orbits of various sizes. While he did not understand why these orbits were stable, he could then assign an energy value to each of the orbits. But the assumption of different sizes for each orbit leads to an interesting consequence. If an electron transitions between levels, by emission or absorption of a photon, it apparently has to do so in an instantaneous "jump." Since it is a particle in the Bohr picture, it cannot exist in any state other than those of the orbits. It cannot move smoothly from one orbit to another, as this seemingly would require it to absorb or radiate energy continuously in values other than that of the photon. Bohr extended this concept to collisions between particles, such as the scattering of an electron by a lattice vibration, the latter of which is also quantized so that a fixed amount of energy has to be exchanged in the event. Bohr asserted that the quantum theory would lead to the particles jumping from their initial

---

[*]K. Hess, K. Michielsen, and H. de Raedt, *Eur. Phys. Lett.* **87**, 60007 (2009).
[†]R. Brady and R. Anderson, arXiv:1305.6822.
[‡]D. K. Ferry, *Fluc. Noise Lett.* **9**, 395 (2010).

state to their final state instantaneously. As the Copenhagen interpretation expanded, it became a requirement that all such transitions between quantized states for particles must involve these non-classical "quantum jumps" during the process.

In the wave theory, this jump was not required, as we will discuss below. The father of the wave theory, Erwin Schrödinger, found the idea of quantum jumps as very distasteful, and contrary to a normal understanding of physics. Indeed, he considered the idea of quantum jumps as a modern counterpart of Ptolemy's epicycles and destined to be consigned to the dustbin of physics.* Yet, the idea has lived on, propagated by the disciples of the Copenhagen hegemony.

The reason for this survival of a quantum jump paradigm was that Heisenberg's matrix mechanics was a particle approach. Thus, in an interaction, one can only state that the particle has the initial or the final properties, but nothing informative about the actual transition itself can be ascertained. In facing this quantum jump, one finds some irrational conclusions, such as the view that it is impossible to catch a particle while it is transitioning, and the actual physical process may be beyond the reach of a theoretical description. If all this seems to be counter-intuitive, it is perhaps, as Eigler said (above), the left-over baggage of maintaining the particle point of view.

However, as we have emphasized in the previous chapters and sections, there is another view of quantum mechanics, and this is the wave picture. Normally, one teaches wave mechanics, rather than matrix mechanics, allegedly because it is easier for the student to comprehend. Perhaps also, there is an undercurrent of feeling that the two approaches are not really equivalent and that only wave mechanics can truly capture the essential physics of not only the ultrasmall (as we discuss in the next chapter), but also the ultrafast. In fact, one can describe the transition between the first state and the second state using the modern power of wave mechanics.

---

*E. Schrödinger, *Br. J. Phil. Sci.* 3, 109 (1952); 3, 233 (1952).

The obvious task in defining a deviation from the idea of a jump is to ask the question of just how long does it take for an interaction and state transition to occur. Classically, the treatment of a collision duration may be tracked back at least as far as Lord Rayleigh in 1906 (although he refers to an earlier discussion in 1881 by Hertz), where he was concerned with the collision between two elastic solid bodies. Quantum mechanically, however we can use the wave approach to compute the scattering probability through what is called the Fermi golden rule, a result for the long time limit of first order, time-dependent perturbation theory. One does not have to take the long time limit, and the buildup of the perturbation can be treated in some detail. The time required to establish the long-time results has been discussed by several people for both electron scattering in condensed matter and for the emission of a photon from an atom.* Let us consider the absorption of a photon in a semiconductor for illustrating the important physics of the process, but will discuss it in terms of the creation of an electron–hole pair in a semiconductor.

In the wave mechanics approach, the idea of absorbing a photon from an incident electromagnetic field, it is important to understand that the electromagnetic field first creates a correlation between the initial and final state. This correlation creates what is called a *polarization* state which involves the wave function for the initial state of an electron in the lower energy state, the valence band, and the wave function for the final state of an electron in the conduction band. This polarization is not a particle itself, so it cannot exist in the particle picture. But this step is the first part of the transition described by perturbation theory. The electromagnetic field can then further interact with this polarization with two results. One is that the correlation is broken up with the electron remaining in the valence band, so that no absorption of the photon occurs. The second possibility is that the electron winds up in the conduction band and the photon is annihilated. The difference in these two possibilities

---

*D. K. Ferry et al., *Phys. Rev. Lett.* **67**, 633 (1991), and the references therein.

is the contribution to what is called the quantum efficiency of light absorption by the semiconductor. The polarization state then exists for only a short time, perhaps a few femtoseconds. Yet, this polarization is a very real quantity, described nicely within the wave mechanical picture. Moreover, as Hartmuth Haug points out in his book, the polarization can be measured experimentally by nonlinear optical techniques such as four wave mixing.* So, clearly, a measurable quantity such as the polarization must be a real process even in quantum mechanics. But this raises a problem with Bohr's complementarity as there is no classical counterpart to this polarization. Even though it is a real quantum mechanical state that exists only for a short time, it does not correspond to a classical particle even though it is experimentally measurable.

But the particle crowd never gives up easily. Thus, there is a description that has been clabbered together so that they can explain the process within the particle view of quantum mechanics. The idea is this: The photon excites the electron, which then oscillates between the conduction band and the valance band. Then, at some point, this excited electron decides whether to stay in the conduction band or to return to the valence band, freeing the photon to go on its merry way. And so, the particle picture lives on, allowing Bohr to rest easy in his grave. But this picture has to have the particle existing in states that are not allowed during its oscillation. It would seem that they want to have their cake and eat it too.

What this, and the previous section, means is that in my mind there is no real reason to accept that Bohr's view of the world is required to believe in quantum mechanics. To be sure, there have been many other interpretations of quantum mechanics, and each such variation has a certain set of disciples. Yet, the so-called Copenhagen interpretation seems to hold a fascination as the dogmatic, or catholic, interpretation, which should be

---

*H. Haug, in *Optical Nonlinearities and Instabilities in Semiconductors*, ed. H. Haug (Academic, San Diego, 1988).

worshipped by all. For sure, Bohr tried to install this view via "proof by intimidation," as the treatment of Schrödinger shows. Perhaps, it is best said by voices from the past. Norman Levitt,* in reviewing James Cushing's book *Quantum Mechanics: Historical Contingency and the Copenhagen Hegemony*, provides this summary in trying to account for the dogmatic Copenhagen interpretation, suggests that part of it might arise from "the role of sheer ambition," but also provides this link to an earlier time as

> by an insight from Francis Bacon (The New Organon, Book I, Aphorism 88): They did it "all for the miserable vainglory of having it believed that whatever has not yet been discovered and comprehended can never be discovered and comprehended hereafter."

Quantum Computing

In an earlier chapter, we talked about the search for dissipation free computers, and their connection to perpetual motion machines. There, we had a delightful character, referred to as Dr. D. It turns out that he followed up his earlier work with an observation that since quantum mechanics was described by a wave equation with no dissipation, we could obviously make a dissipation free quantum computer. Since my group had taken pains to try to demonstrate that the search for a dissipation free computer was contrary to most physics and thermodynamics, we naturally were sucked into the discussion on the efficacy of quantum computers.

A classical computer uses individual switches which are either "on" or "off. In one state, they represent a logical 1, while in the other state they represent a logical 0, where these two values are the only possible values in binary logic. Transistors

---

*N. Levitt, *Phys. Today* **48**(11), 84 (1995).

work well in this scenario, since it is easy to ascertain whether they are on (carrying current) or off (not carrying current). Each switch corresponds to one bit in the binary word (such as 64 bits on a modern microprocessor). The primary difference between the bits on a classical digital computer and the so-called qubits (short for quantum bit) of a quantum computer is that the latter incorporate quantum mechanical phase factors that allow a continuous range of projection onto the "0" and "1" states. As such, the qubits themselves should be thought of as analog objects, rather than digital or binary objects. That is, the state is a continuous (complex) variable, basically the phase of the complex qubit, instead of merely a 0 or a 1. Yet, a quantum computer, like a classical computer, is a set of interconnected processing elements, but now the latter are a set of qubits. In general, the speed of a quantum computer is likely no faster than that of a classical digital computer, which processes instructions in a sequential stream. But it is hoped that quantum mechanics promises more, and different quantum states can be *entangled* on a single set of qubits. The entanglement leads to a single wave function describing the many partners, and our qubit can work on this single wave function. Thus, the argument goes that this allows many "processes" to be done in parallel. It is through this inherent parallelism that quantum computers are thought to gain a speed advantage over the classical computer. Whether or not this is the case in the real world is the subject of some debate these days.

Let us return to the issue of whether or not these machines can be dissipation free. It is true that the Schrödinger equation is time reversible, but this is the textbook case and only applies to a *closed* system. But this can't be applied to a computer, since it has to be an *open* system—we need to feed information into the computer and we hopefully get answers from the computer. And, as was discussed earlier, computation consists of a series of steps, each of which forces the machine from one state to the next. By necessity, the computer has a single direction as time progresses, and this means that dissipation must occur to keep the machine from thermalizing, where all information will be lost. If we think of a simple quantum gate, there will be a state of the system before the gate and another state of the system after the gate. The process of the gate's operation entails this preferred direction of time, and the irreversibility here is usually coupled to dissipation. Moreover, the gate itself is usually physically laid out over a small area, and the processing gives us what we could call a flow of information from one side to the other. In doing so, there is an inherent velocity (or momentum) which accounts for a certain distance across the gate being traversed in a small amount of time. In the qubit gate, though the change of state can only be accounted for by a phase change of the wave function describing the qubit. But the velocity, or momentum, actually resides in the phase of the wave function as an appropriate parameter. So, we have a change of position from one side of the gate to the other, and this means that there must be a variability in the velocity or momentum that has to satisfy the Heisenberg uncertainly relation. Since we don't know the velocity exactly, we cannot say for sure just how long it will take for the quantum operation to be finished. As a result, there will be some indeterminacy in the exact specification of the output state of the gate. This corresponds to a possible error, which some would think results from noise in the system. But it is an uncertainty induced error.

In the classical gate, we correct for errors by using the nonlinearity of the transistor transfer relationship. In essence, we ensure that the transistor being "off" resets the voltage to a set standard level (usually the supply voltage), and this corresponds to the logical 0 or 1. Similarly, the transistor being "on" resets the voltage to a different level (usually the ground, or zero, level). With the quantum gate, we cannot do this. The fact that the qubit is analog in nature implies that we must try to keep the gate linear, and there is no way to reset the levels to the voltages as that would defeat the basic premise of the qubit. So the modern view in quantum computing is that there is going to be some inherent dissipation and loss of coherence of the phase that is important in the wave function. We must find methods to minimize the loss of this phase information, either through trying to build the gate from physical processes with long coherency times, or to use a form of error correction. The later is commonly used in classical computers to try to avoid bit errors which can arise from bad gates, but for the quantum computer we need quantum error correction, and this has become something of a hot topic in quantum information processing.

Nevertheless, there continues to be a great deal of effort applied to bringing quantum computing into some state of reality. This involves both basic research into the physics of how a quantum gate can be created as well as into the fundamentals of the computing process. As with many fields of scientific investigation, the participating scientists view the future as being bright (otherwise, why would they participate—nobody likes to work on something that is certain to fail). But there are the naysayers. Some of these ask very real questions about just how useful these machines may be, and whether that promise can ever be fulfilled.*† These views raise questions about the efficacy of the technology, but are not solid barriers. Rather, they describe the natural barriers that arise to all new technologies, and, while they describe daunting challenges, only time and effort will decide how real these barriers are.

Life as a Heretic

As one may suspect in reading about my objections to what seems to be mainstream thought, holding such views can lead to repercussions. In essence, when you hold heretical views, which are contrary to various hegemonies, you will be branded as a heretic. I suppose that I have had this tendency ever since youth, but I am continuously surprised that such ideas can continue to

---

*J. Gea-Banacloche and L. B. Kish, *Proc. IEEE* **93**, 1858 (2005).
†M. I. Dyakonov, *JETP Lett.* **98**, 514 (2013).

hold immense power over a great fraction of the scientists even today. But I suppose this is the face of politics, since when these differences of viewpoint exist between leaders of countries, or even of factions or clans, violence can ensue. One might think that this would not carry over to science, where the various fields are inhabited by quite well educated and smart people. This, however, would be incredibly naïve on the part of anyone thinking that politics doesn't play a role in science. To begin with, the funding of science is determined by politicians in government. These people are likely not to be knowledgeable in the fields of science. Then, it certainly plays a dramatic role in the supposedly blind peer review by which various proposals are evaluated. And, of course, it plays a role within the universities where the work is done. Some insightful person once suggested that politics at a university are so intense because there is so little to gain.

When one lives as a heretic, there is a natural attraction to other heretics. Perhaps we enjoy sharing our own experiences of frustration with an understanding peer group. But it is a fact that we are drawn together. One might even think that we are an underground movement seeking to usurp the rightful power of the self-proclaimed hegemonies against which we rail. The problem is that we are not organized to the level of a true underground movement that has any chance of succeeding. Even if we were, I suspect that the impact would not be terribly great. While the fields of science are populated with people who in general are far better educated than the population at large, the sociology within the fields of science is not much different than that of the population at large. Just as a significant fraction of the population is not tuned into the economic and political questions of the day, the typical scientist is just not aware of the great questions that still exist on the philosophical foundations of many fields of science. As the former group usually neglects to vote, the latter group just doesn't worry about these questions, and enjoys status within the great herd of acceptors of the faith, as promulgated by the ruling hegemonies. While the herd can move from special topic to the next special topic of the day, it never wavers in its acceptance of the political correctness of their scientific view.

This is a sad description of the state of the fields, but one cannot spend time worrying about it. If you do spend time fighting against the doctrines, it could make you an unhappy camper. Eventually, you have to reach a state of acceptance, not of the doctrines, but of the likelihood that you cannot affect any change in a realistic manner. Instead, one has to spend their limited time trying to work on the problems for which they may be able to provide solutions. And, with these successes, one may be able to gain a level of recognition. With recognition, perhaps others may pay more attention to the views for which we have been branded as heretics. With continued small steps, perhaps the day will arrive when these strange views promulgated by the hegemonies will finally be consigned to the dustbin.

Chapter 9

How Big Is an Electron

In this last chapter, I want to describe a thread that began just under two decades ago. At the same time, the reader may wonder why I put so much effort into the last chapter with regard to the manner of understanding quantum mechanics. The importance of this lies in the thread to be discussed here, which connects us back to the very small transistors that have arisen as Moore' law has continued to progress. Today, we utilize individual transistors in our chips that may be described as nanoscale devices. Critical lengths in these devices are at the 20 nm (or smaller) scale. But in these devices we expect that certain quantum effects will dominate the transport, and therefore the performance of the devices. In these small structures, one must begin to worry about the effective size of the electrons (or holes) themselves. Of course, there are many estimates about the effective size, most of which would preclude having even a single electron in some of the devices being made today. Certainly, the effective size is going to be larger than the very small estimate from classical physics. Moreover, how can we connect the larger size with the apparent classical behavior of the devices.

50 Years in the Semiconductor Underground
David K. Ferry
Copyright © 2015 Pan Stanford Publishing Pte. Ltd.
ISBN 978-981-4613-34-7 (Hardcover), 978-981-4613-35-4 (eBook)
www.panstanford.com

Consider, for example, the most obvious definition of the size that arises from considering a wave packet description of the electron. This would be the de Broglie wavelength. For an electron in silicon, at room temperature, the thermal de Broglie wavelength is about 8–18 nm. So, one can see that there is a concern about fitting a wave packet of this size into a device that is only 20 nm in size. Such a wave packet would begin leaving the channel of the device before it had completely entered the channel. In truth, the electron may actually be larger. In today's transistors, the electrons are pushed into a potential well that exists under the oxide gate insulator. Squeezing the electron in this one dimension means that it will expand in the other two dimensions, so its length along the channel may actually be larger than the thermal de Broglie wavelength. On the other hand, considerations on the role of heavy scattering in the contact regions of the device suggest that the wave packet may only be about half the thermal de Broglie wavelength. Thus, a reasonable compromise would suggest that a wave packet of a few nanometers is appropriate for the electron wave packet.

It is clear from the preceding chapter that an approach based upon the wave function view of quantum mechanics is the correct one to take. First, the remarks of Don Eigler tell us that we will always get the right answer with this approach. Secondly, we don't have to worry about the electron jumping from one state to the next, as there is a very smooth transition in the wave description. Thirdly, the philosophical interpretations of the device and performance are much easier to live with within this approach. In a sense, reality and causality will still be with us, with no magical occurrences arising from an ill-posed observer.

Focusing In on the Size

In actually trying to conceive of just how big the electron should be in the quantum world, there are several factors to be considered. Perhaps the most important one is the fact that the single electron does not exist alone, but is surrounded by a large number of other electrons. So, when we talk about *an* electron, we are really talking about a member of a large group of electrons. This leads to the need to describe the wave packet in the presence of these other electrons. Most considerations in introductory physics treat the electron via a plane wave, but such a wave is infinite in extent, as it has only a single momentum state. For our electron, we have to consider the totality of the group of electrons, which compose a large range of such plane wave states. These are the occupied momentum states for our ensemble of electrons, which sometimes are called an electron gas. To describe the wave packet for one of our electrons, we account for all of the occupied momentum states by summing over them all and then Fourier transforming to get the packet in real space. There are a couple of ways to describe the set of momentum states.

First, let us consider the electrons that were trapped in Don Eigler's quantum corral, that we discussed earlier. These can be considered to be members of a two-dimensional ensemble of electrons, as they were at the surface of the metal. Hence, all states up to the Fermi energy are occupied, so that the momentum states are all of those whose momentum is below that corresponding to the Fermi energy. In this two-dimensional world, the Fermi momentum varies as the square root of the surface electron concentration. Thus, for a surface electron concentration of about 10^{13} cm^{-2}, we would find a Fermi wave vector of just under 8×10^7 cm^{-1}. Now, when we take this uniform distribution of momentum states, the Fourier transform gives us a distinct special function, which is a first order Bessel function of the first kind divided by the radial distance. The Bessel function is a function of the radial distance times the Fermi momentum wave vector. This is shown in the attached figure. The expected oscillations of the Bessel function are damped by the rapid decay of the overall wave packet. It turns out that this

The wave packet corresponding to an electron in the degenerate Fermi gas.

is similar to the shape that is seen in Eigler's images. Now, this function has a central width which is about π, so that we find that twice the radius of the wave packet is about the de Broglie wave length for a momentum state at the Fermi energy. This gives a value of approximately 5 nm. However, this is about five times larger than the size measured in the experiments.* One reason for the smaller experimental size could be the confinement of the electron within the corral, which would be expected to lead to a smaller wave packet due to repulsion from the walls.

While the surface electron density was estimated for the copper material that was used in the experiment, similar densities can be reached in the Si MOSFET used in today's integrated circuits. But a more usual value at low bias is a density of the order of 10^{12} cm^{-2}. Then, the Fermi wave vector is a few times 10^6 cm^{-1}, and this leads to a wave packet size of some 12 nm, which is larger than our earlier estimate. On the other hand, the distribution of momentum states at this density, and room temperature, is closer to a classical one, instead of the fully quantum one considered above. That is, the actual distribution of the momentum states is a Gaussian one, and this gives rise to a value that is about half the de Broglie thermal wave length, or about 6 nm. One may notice that the wave packet shown in the figure discussed above appears quite close to a Gaussian, so that this approximation may well be considered to be appropriate. This now tends to agree with the

---

*M. F. Crommie, C. P. Lutz, and D. M. Eigler, *Science* **262**, 218 (1993).

surface density value of the previous paragraph. That is, we can expect the wave packet for an electron in a semiconductor to be of the order of 5–6 nm in extent, but slightly smaller if confined in a quantum structure.

The above discussion tells us that the effective size of the electron depends upon either the temperature, in the classical limit, or the density, in the full quantum limit. Within a modern device, the density and temperature vary throughout the device (the latter is actually above the ambient temperature due to the power dissipation). Thus, the effective wave packet size will also vary throughout the device. But because of confinement within the device, the effective size will be modified by its interaction with the environment and walls of the structure, just as seen in the IBM experiments of the quantum corral. But, in any of these scenarios, we are talking about a size on the scale of a few nanometers. This is considerably larger than the classical idea of the size of an electron. When the device size is only a few 10s of nanometers, it will be necessary to consider the extended nature of the electron in any discussion of the transport physics of the device. What effect will this play in our understanding of how such small devices work?

Why Is It Still Classical

There is one thing that is abundantly clear from the success of the industry in making very small silicon transistors. These devices act quite normally, and behave in a manner nearly identical to the way they did with much larger sizes. Part of this is, of course, the success of process development which aims to have this exact result occur. But the take-home message is that quantum effects have not made any major impact that is observable to the casual user of the devices and chips. How can this be, when the size of the electron was found above to be a few nanometers. Certainly one might expect that this would impact device performance. But the real observations suggest otherwise.

To understand how this dilemma can be resolved, we have to return to our understanding of how density and potentials enter into the quantum mechanical description of the device performance. In any such device, there must be a conservation of total energy. Energy supplied to the device by external voltages and currents must go somewhere, and this is usually into heat that flows out of the device to a reservoir. In describing this, one can write down an equation for the total energy, in which one term is the role-played by the local potential within the device. This latter is the self-consistent potential found by solving Poisson's equation subject to the device geometry. This term involves this local potential weighted by the electron density at that position. That is, the total term may be viewed as a sum over the individual electrons weighted by the local potential at the location of each electron. If each electron is described by a wave packet, we may think of this as a shape, like the Gaussian, centered around what could be considered to be the classical location of the electron. If we take the classical location of the electron as a point in space, we describe that as a delta function of the general coordinates—this function may be thought of as being zero everywhere except the single point at which the electron exists. The corresponding wave packet is then a general shape, like the Gaussian envelope, centered upon this single point in space. The summation described for the total energy is now a sum over the local potential, the wave packet shape, and the sites of the electrons. Mathematically, in any such summation or integral, we can rearrange the terms in various manners. The important one for us is to move the wave packet shape from the electron to the potential. This now creates a smoothing of the original potential and gives us what can be called an *effective* potential. So, if we think about the electron as a classical particle, then the potential is a smoothed version of the normal self-consistent potential. In a sense, this reflects the fact that quantum mechanics in general prefers soft potentials; we have to work hard to impose sharp potential variations. If we allow self-consistency, these sharp edges are smoothed. It is this principle that allows us to continue to think about the individual electrons as point particles, if we account for the smoothing of the potential by quantum forces, which is just the process described above.

The approach described above creates what is called an *effective potential*. The ideas are even outlined in Feynman and Hibbs' textbook.* In applying the approach to the simulation of semiconductor devices, it joins a number of other attempts to incorporate quantum effects into the devices. But the approach has been shown to give the proper corrections to the capacitance in a typical MOS transistor. Using the effective potential pushes the charge slightly away from the oxide interface, just as occurs for normal quantization. And, it raises the minimum energy in a manner consistent with the quantization. The results obtained with this approach seem to agree well with those obtained by direct solution of the Schrödinger equation along with the Poisson's equation for the self-consistent potential.† We have used this approach in direct simulations of a number of semiconductor devices over the subsequent years, and find that it gives good results. Of course there will be some limitations and it doesn't work so well in environments where strong confinement quantization arises.

But it is quite clear from this approach that the role of quantum mechanics causes well-defined, predictable changes to the potential structure within the device. These effects are clearly causal in nature, and the changes in the device properties are observable and reliable. There is little probability in the measurement outcomes, at least from these effects. And so, our device appears to retain its classical behavior, even when appearing at such a small size.

---

*R. P. Feynman and A. R. Hibbs, *Quantum Mechanics and Path Integrals* (McGraw-Hill, New York, 1965).
†D. K. Ferry, *VLSI Design* **13**, 155 (2001).

About the Author

David Ferry is Regents' Professor in the School of Electrical, Computer, and Energy Engineering at Arizona State University. He is also graduate faculty in the Department of Physics and the Materials Science and Engineering program at ASU, as well as visiting professor at Chiba University in Japan. He came to ASU in 1983 following shorter stints at Texas Tech University, the Office of Naval Research, and Colorado State University. In the distant past, he received his doctorate from the University of Texas, Austin, and spent a postdoctoral period at the University of Vienna, Austria. He enjoys teaching (which he refers to as "warping young minds") and research. The latter is focused on semiconductors, particularly as they apply to nanotechnology and integrated circuits, as well as quantum effects in devices. In 1999, he received the Cledo Brunetti Award from the Institute of Electrical and Electronics Engineers, and is a Fellow of this group as well as the American Physical Society and the Institute of Physics, UK. He has been a Tennessee Squire since 1971 and an admiral in the Texas Navy since 1973.

Index